Computer-aided Detection of Architectural Distortion in Prior Mammograms of Interval Cancer

Synthesis Lectures on Biomedical Engineering

Editor
John D. Enderle, *University of Connecticut*

Lectures in Biomedical Engineering will be comprised of 75- to 150-page publications on advanced and state-of-the-art topics that span the field of biomedical engineering, from the atom and molecule to large diagnostic equipment. Each lecture covers, for that topic, the fundamental principles in a unified manner, develops underlying concepts needed for sequential material, and progresses to more advanced topics. Computer software and multimedia, when appropriate and available, are included for simulation, computation, visualization and design. The authors selected to write the lectures are leading experts on the subject who have extensive background in theory, application and design.

The series is designed to meet the demands of the 21st century technology and the rapid advancements in the all-encompassing field of biomedical engineering that includes biochemical processes, biomaterials, biomechanics, bioinstrumentation, physiological modeling, biosignal processing, bioinformatics, biocomplexity, medical and molecular imaging, rehabilitation engineering, biomimetic nano-electrokinetics, biosensors, biotechnology, clinical engineering, biomedical devices, drug discovery and delivery systems, tissue engineering, proteomics, functional genomics, and molecular and cellular engineering.

Computer-aided Detection of Architectural Distortion in Prior Mammograms of Interval Cancer
Shantanu Banik, Rangaraj M. Rangayyan, and J.E. Leo Desautels
2012

Chronobioengineering: Introduction to Biological Rhythms with Applications, Volume 1
Donald McEachron
2012

Medical Equipment Maintenance: Management and Oversight
Binseng Wang
2012

Fractal Analysis of Breast Masses in Mammograms
Thanh M. Cabral and Rangaraj M. Rangayyan
2012

Fundamentals of Respiratory Sounds and Analysis
Zahra Moussavi
2006

Advanced Probability Theory for Biomedical Engineers
John D. Enderle, David C. Farden, and Daniel J. Krause
2006

Intermediate Probability Theory for Biomedical Engineers
John D. Enderle, David C. Farden, and Daniel J. Krause
2006

Basic Probability Theory for Biomedical Engineers
John D. Enderle, David C. Farden, and Daniel J. Krause
2006

Sensory Organ Replacement and Repair
Gerald E. Miller
2006

Artificial Organs
Gerald E. Miller
2006

Signal Processing of Random Physiological Signals
Charles S. Lessard
2006

Image and Signal Processing for Networked E-Health Applications
Ilias G. Maglogiannis, Kostas Karpouzis, and Manolis Wallace
2006

Three-dimensional Image Processing Techniques to Perform Landmarking and Segmentation to Computed Tomographic Images
Shantanu Banik, Rangaraj M. Rangayyan, and J.E. Leo Desautels

ISBN: 978-3-031-00528-2   paperback
ISBN: 978-3-031-01656-1   ebook

DOI: 10.1007/978-3-031-01656-1

A Publication in the Springer series
SYNTHESIS LECTURES ON BIOMEDICAL ENGINEERING

Lecture #47
Series Editor: John D. Enderle, *University of Connecticut*
Series ISSN
Synthesis Lectures on Biomedical Engineering
Print 1930-0328   Electronic 1930-0336

Computer-aided Detection of Architectural Distortion in Prior Mammograms of Interval Cancer

Shantanu Banik, Rangaraj M. Rangayyan, and J.E. Leo Desautels

University of Calgary, Calgary, Alberta, Canada

SYNTHESIS LECTURES ON BIOMEDICAL ENGINEERING #47

ABSTRACT

Architectural distortion is an important and early sign of breast cancer, but because of its subtlety, it is a common cause of false-negative findings on screening mammograms. Screening mammograms obtained prior to the detection of cancer could contain subtle signs of early stages of breast cancer, in particular, architectural distortion. This book presents image processing and pattern recognition techniques to detect architectural distortion in prior mammograms of interval-cancer cases. The methods are based upon Gabor filters, phase portrait analysis, procedures for the analysis of the angular spread of power, fractal analysis, Laws' texture energy measures derived from geometrically transformed regions of interest (ROIs), and Haralick's texture features. With Gabor filters and phase-portrait analysis, 4,224 ROIs were automatically obtained from 106 prior mammograms of 56 interval-cancer cases, including 301 true-positive ROIs related to architectural distortion, and from 52 mammograms of 13 normal cases. For each ROI, the fractal dimension, the entropy of the angular spread of power, 10 Laws' texture energy measures, and Haralick's 14 texture features were computed. The areas under the receiver operating characteristic (ROC) curves obtained using the features selected by stepwise logistic regression and the leave-one-image-out method are 0.77 with the Bayesian classifier, 0.76 with Fisher linear discriminant analysis, and 0.79 with a neural network classifier. Free-response ROC analysis indicated sensitivities of 0.80 and 0.90 at 5.7 and 8.8 false positives (FPs) per image, respectively, with the Bayesian classifier and the leave-one-image-out method. The present study has demonstrated the ability to detect early signs of breast cancer 15 months ahead of the time of clinical diagnosis, on the average, for interval-cancer cases, with a sensitivity of 0.8 at 5.7 FP/image. The presented computer-aided detection techniques, dedicated to accurate detection and localization of architectural distortion, could lead to efficient detection of early and subtle signs of breast cancer at pre-mass-formation stages.

KEYWORDS

angular spread of power, architectural distortion, breast cancer, coherence, computer-aided diagnosis (CAD), entropy, feature selection, fractal dimension, Gabor filters, Haralick's texture measures, interval cancer, Laws' texture energy measures, mammography, orientation strength, pattern recognition, phase-portrait analysis, prior mammograms, texture analysis

Shantanu Banik dedicates this book to his daughter

Swasti Shuchismita Banik

Contents

Preface

This book presents image processing and pattern recognition techniques to detect architectural distortion in prior mammograms of interval-cancer cases. The methods are based upon Gabor filters, phase-portrait analysis, procedures for the analysis of the angular spread of power, fractal analysis, Laws' texture energy measures derived from geometrically transformed ROIs, and Haralick's texture features. An extensive review of the state of the art in CAD of architectural distortion in mammography is presented; it is evident from the review that there is a need and scope for further developments in the detection of architectural distortion.

The methods developed for the detection of architectural distortion are based on the oriented textural patterns present in mammographic images that are extracted by using a bank of real Gabor filters. Linear phase-portrait modeling is used for the detection of potential sites of architectural distortion in prior mammograms. In addition, fractal analysis, analysis of the angular spread of power using higher order entropy measures, structured pattern analysis using Laws' texture energy measures, and Haralick's methods for statistical analysis of texture are used for the detection of architectural distortion in prior mammograms of interval-cancer cases. Although the use of several types of features is presented, an efficient combination of different types of selected features could be used for the detection and characterization of architectural distortion.

Computer-aided detection of architectural distortion has not been studied adequately: there is increasing interest in this area at present, as indicated by the appearance of a small number of publications addressing parts of the problem. Moreover, computer-aided detection of architectural distortion in prior mammograms, in particular, of interval-cancer cases, is an important but yet-to-be addressed field to facilitate early detection of breast cancer. The performance of the methods developed in this work for the detection of architectural distortion in prior mammograms of interval-cancer cases resulted in a sensitivity of 80% at 5.7 FP/image; the result compares favorably with the performance of other methods reported in the scientific literature for the detection of spiculated masses, which is a simpler problem.

This book is intended for engineering students and professionals working on medical images and on the development of systems for computer-aided diagnosis (CAD). The methods and technical details presented in this book are at a fairly high level of sophistication, and should find use in image processing and pattern recognition for practical applications.

Acknowledgments

This book is the result of the research conducted with support from many individuals and organizations. We gratefully acknowledge Dr. Fábio J. Ayres and Ms. Shormistha Prajna for their contributions to preliminary stages of this research. We thank the Natural Sciences and Engineering Research Council (NSERC) of Canada, the Alberta Heritage Foundation for Medical Research (AHFMR), and the Canadian Breast Cancer Foundation (CBCF): Prairies/NWT Chapter for research grants provided.

We gratefully acknowledge the financial support provided by the Department of Electrical and Computer Engineering, J. B. Hyne Research Innovation Award, Robert B. Paugh Memorial Scholarship in Engineering, Queen Elizabeth II Graduate (Doctoral) Scholarship, University of Calgary Graduate Faculty Council Scholarship — Doctoral, University Technologies International Inc. (UTI) Fellowship, University of Calgary Alumni Association Graduate Scholarship, and the University of Calgary.

We thank Dr. D. P. Chakraborty, University of Pittsburgh, for providing the JAFROC software package and for his assistance with the same. We also thank Dr. Mauro Tambasco for providing a program to synthesize images with specified fractal dimension values.

We thank the Institute of Electrical and Electronics Engineers (IEEE), the IEEE Engineering in Medicine and Biology Society (EMBS), the International Society for Optics and Photonics (SPIE), the Society of Imaging Informatics in Medicine (SIIM), International Journal of Computer Assisted Radiology and Surgery (IJCARS), and Springer Science+Business Media for their permission to use published material.

Some of the materials and illustrations have been reproduced, with permission from the associated organizations, from our publications as are listed below:

1. S. Banik, R. M. Rangayyan, and J. E. L. Desautels. "Measures of Angular Spread and Entropy for the Detection of Architectural Distortion in Prior Mammograms." *International Journal of Computer Assisted Radiology and Surgery*, vol. 8(1), pp. 121–134, January 2013. © Springer.

2. S. Banik, R. M. Rangayyan, and J. E. L. Desautels. "Detection of architectural distortion using statistical measures of correlation and stationarity." *Rendiconti del Circolo Matematico di Palermo*, accepted February 2012. In press. © Springer.

3. S. Banik, R. M. Rangayyan, and J. E. L. Desautels. "Detection of Architectural Distortion in Prior Mammograms." *IEEE Transactions on Medical Imaging*, vol. 30(2), pp. 279–294, February 2011. © IEEE.

4. R. M. Rangayyan, S. Banik, and J. E. L. Desautels. "Computer-aided Detection of Architectural Distortion in Prior Mammograms of Interval Cancer." *Journal of Digital Imaging*, vol. 23(5), pp. 611–631, October 2010. © Springer, SIIM.

5. R. M. Rangayyan, S. Banik, and J. E. L. Desautels. "Detection of Architectural Distortion in Prior Mammograms using Statistical Measures of Angular Spread." In J. S. Suri, Y. K. Ng, and R. M. Rangayyan, Eds., *Diagnostic and Therapeutic Applications of Breast Imaging*, vol. II, SPIE Press, Bellingham, WA, 2013. In press. © SPIE.

6. S. Banik, R. M. Rangayyan, and J. E. L. Desautels. "Image Processing and Pattern Classification Techniques for the Detection of Architectural Distortion in Prior Mammograms of Interval-cancer Cases." In J. S. Suri, S. V. Sree, K.-H. Ng, and R. M. Rangayyan, Eds., *Diagnostic and Therapeutic Applications of Breast Imaging*, pp. 197–242, SPIE Press, Bellingham, WA, February 2012. © SPIE.

7. S. Banik, R. M. Rangayyan, and J. E. L. Desautels. "Digital Image Processing and Machine Learning Techniques for the Detection of Architectural Distortion in Prior Mammograms." In K. Suzuki, Ed., *Machine Learning in Computer-aided Diagnosis: Medical Imaging Intelligence and Analysis*, pp. 24–63, IGI Global, Hershey, PA, January 2012. © IGI Global.

8. R. M. Rangayyan, S. Banik, and J. E. L. Desautels. "Statistical Measures of Correlation and Stationarity for the Detection of Architectural Distortion in Prior Mammograms of Interval-cancer Cases." In *Proceedings of the 26th International Congress and Exhibition: Computer Assisted Radiology and Surgery*, Special session on breast CAD, pp. 1–2, Pisa, Italy, June 2012. © Springer.

9. R. M. Rangayyan, S. Banik, and J. E. L. Desautels. "Detection of Architectural Distortion in Prior Mammograms Using Measures of Angular Dispersion." In the *2012 IEEE International Symposium on Medical Measurements and Applications (MeMeA 2012)*, pp. 87–90, Budapest, Hungary, May 2012. © IEEE.

10. S. Banik, R. M. Rangayyan, and J. E. L. Desautels. "Detection of Architectural Distortion in Prior Mammograms of Interval Cancer Using Measures of Angular Spread and Tsallis Entropy." In *Proceedings of the 25th International Congress and Exhibition: Computer Assisted Radiology and Surgery*, vol. 6(1), pp. S188–S189, Berlin, Germany, June 2011. © Springer.

11. S. Banik, R. M. Rangayyan, and J. E. L. Desautels. "Rényi Entropy of Angular Spread for Detection of Architectural Distortion in Prior Mammograms." In *Proceedings of the 2011 IEEE International Symposium on Medical Measurements and Applications (MeMeA 2011)*, pp. 609–612, Bari, Italy, May 2011. © IEEE.

12. R. M. Rangayyan, S. Banik and J. E. L. Desautels. "Detection of Architectural Distortion in Prior Mammograms Using Measures of Angular Distribution." In *Proceedings of SPIE*

Medical Imaging 2011: Computer Aided Diagnosis, R. M. Summers and B. van Ginneken, Eds., vol. 7963, pp. 796308: 1–9. Orlando, FL, February 2011. © SPIE.

13. S. Banik, R. M. Rangayyan, and J. E. L. Desautels. "Detection of Architectural Distortion in Prior Mammograms of Interval Cancer Using Laws' Texture Energy Measures." In *Proceedings of the 24th International Congress and Exhibition: Computer Assisted Radiology and Surgery*, vol. 5(1), pp. S200–S201, Geneva, Switzerland, June 2010. © Springer.

14. S. Banik, R. M. Rangayyan, and J. E. L. Desautels. "Detection of Architectural Distortion in Prior Mammograms Using Fractal Analysis and Angular Spread of Power." In *Proceedings of SPIE Medical Imaging 2010: Computer Aided Diagnosis*, N. Karssemeijer and R. M. Summers, Eds., vol. 7624, pp. 762408: 1–9, San Diego, CA, February 2010. © SPIE.

15. S. Banik, R. M. Rangayyan, and J. E. L. Desautels. "Detection of Architectural Distortion in Prior Mammograms of Interval-cancer Cases with Neural Networks." In *Proceedings of the 31st Annual International Conference of the IEEE Engineering in Medicine and Biology Society*, pp. 6667–6670, Minneapolis, MN, September 2009. © IEEE.

16. R. M. Rangayyan, S. Banik, S. Prajna, and J. E. L. Desautels. "Detection of Architectural Distortion in Prior Mammograms of Interval-cancer Cases." In *Proceedings of the 23rd International Congress and Exhibition: Computer Assisted Radiology and Surgery*, vol. 4(1), pp. S171–S173, Berlin, Germany, June 2009. © Springer.

Finally, we thank our families for their good wishes, inspiration, patience, appreciation, and loving care.

Shantanu Banik
Rangaraj M. Rangayyan
J. E. Leo Desautels
Calgary, Alberta, Canada
January, 2013.

List of Symbols and Abbreviations

ANN	Artificial neural network
AUC	Area under the ROC curve
A_z	Area under the ROC curve with binormal distribution
BI-RADS	Breast Imaging–Reporting and Data System
CAD	Computer-aided diagnosis
CC	Craniocaudal
CLS	Curvilinear structure
DC	Direct current or zero frequency
FACA	Fractal analysis by circular average
fBM	Fractional Brownian motion
FD	Fractal dimension
FLDA	Fisher linear discriminant analysis
FN	Number of false negatives
FNR	False-negative ratio
FP	Number of false positives
FPR	False-positive ratio
FROC	Free-response receiver operating characteristics
GCM	Gray-level cooccurrence matrix
K	Number of orientations in the Gabor filter bank, spanning the interval $[-\pi/2, \pi/2]$
l	Elongation factor of the real Gabor filter
LOO	Leave-one-out
MIAS	Mammographic Image Analysis Society
MLO	Mediolateral oblique
N	Number of negatives
P	Number of positives
PCA	Principal component analysis
PDF	Probability density function
PPV	Positive-predictive value
PSD	Power spectral density
QDA	Quadratic discriminant analysis
RBF	Radial basis function

RBST	Rubber-band straightening transform
ROC	Receiver operating characteristics
ROI	Region of interest
SD	Standard deviation
SHL	Single-hidden-layer backpropagation neural network
SLFF	Single-layer feedforward neural network
SVM	Support vector machine
TN	Number of true negatives
TNR	True-negative ratio
TP	Number of true positives
TPR	True-positive ratio
τ	Full-width at half-maximum of the Gaussian term in the real Gabor filter

CHAPTER 1

Introduction

1.1 BREAST CANCER AND MAMMOGRAPHY

Breast cancer is one of the major health issues affecting women: it is the most common (aside from nonmelanoma skin cancer) and the second leading cause of cancer related death among women [1, 2]. Breast cancer has become one of the major health problems for women in developed as well as developing countries over the past 50 years [2]. According to the Canadian Cancer Society [3], the lifetime probability of developing breast cancer is one in nine, and the lifetime probability of death due to breast cancer is one in 28. Breast cancer is the most common cancer among Canadian women (excluding nonmelanoma skin cancer), with 1.02% of all women living with the disease [3]. In the year 2010 in Canada, breast cancer accounted for 28% of all new cancers and 15% of cancer-related deaths. In other words, an estimated 23,200 women were diagnosed with breast cancer and 5,300 died due to the same in Canada in the year of 2010; on average, 445 Canadian women were diagnosed with breast cancer and 100 women died due to the same every week in 2010 [3].

Due to the fact that only localized cancer is considered to be treatable and curable, early detection of breast cancer is of utmost importance [4]. Efficient detection of breast cancer in its early stages can play an important role in reducing the associated mortality rates: localized cancer leads to a five-year survival rate of 97.5%, whereas cancer that has spread to distant organs has a five-year survival rate of only 20.4% [5].

The rate of incidence or detection of breast cancer increased steadily from 1980 to the early 1990s, mostly due to the increased number of mammographic screening programs. Death rates due to breast cancer have declined in every age group since at least the mid 1980s [3]. Breast self-examination has been found to be inadequate for accurate and early detection of breast cancer: there is no significant evidence of a reduction in the mortality rate from breast cancer among women who perform breast self-examination regularly, compared to those who do not [6, 7].

Mammography is, at present, the most widely available screening tool for the early detection of breast cancer [6]. It can demonstrate pronounced signs of abnormality, such as masses and calcifications, as well as subtle signs, such as bilateral asymmetry and architectural distortion [8]. Mammographic screening has been shown to be effective in reducing mortality rates due to breast cancer: screening programs have reduced mortality rates by 30% to 70% [4, 5]. Accurate detection of malignancies via mammographic screening could result in a reduction in the number of benign biopsies, while maintaining the desired sensitivity, which could not only reduce costs associated with biopsies but also lessen patient suffering caused by the traumatic experience of biopsy [9].

In a study on evaluating the importance of mammographic screening programs, Cady and Chung [10] highlighted the reduction in mortality rate due to several screening programs in Sweden, the Netherlands, the United Kingdom, Finland, Italy, and the United States. The major drawbacks of screening have been found to be: higher incidence of unnecessary (benign) biopsies, cost and quality of interpretation of mammograms versus the experience of the radiologists, and the psychological consequences of errors, such as the anxiety caused by a false-positive (FP) result and the improper reassurance provided by a false-negative (FN) test. The authors concluded that, despite having some drawbacks, the practice of mammographic screening must be encouraged and expanded.

Screening mammography is recommended for asymptomatic women and involves two views of each breast: the craniocaudal (CC) view and the mediolateral oblique (MLO) view [11]. Figure 1.1 shows a pair of CC and MLO views of a normal breast; the pectoral muscle is visible in the MLO view. The Canadian Cancer Society recommends women from 50 to 69 years of age to participate in regular mammographic screening once every two years.

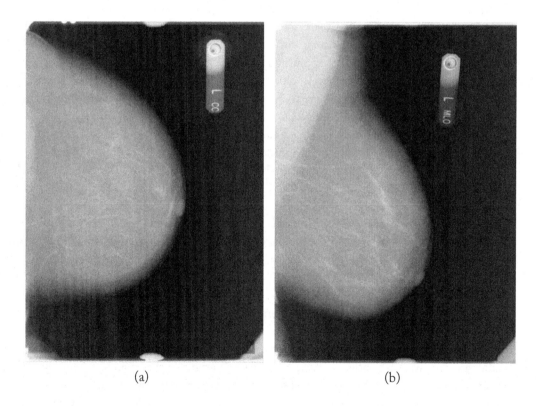

(a)   (b)

Figure 1.1: (a) The craniocaudal (CC) view and (b) the mediolateral oblique (MLO) view of a normal breast.

1.2   MAMMOGRAPHIC SIGNS OF BREAST CANCER

Mammography is an effective noninvasive imaging technique, based on the use of low-dose X rays, to examine the breast [12]. On a mammogram, the various levels of gray correspond to the densities of the different tissues being imaged: lighter or brighter appearance of a region is associated with denser tissues as more X rays are absorbed or attenuated. Traditional mammography uses a screen-film combination to capture the image, and the image is produced after the film is processed. Modern diagnostic centers are equipped with full-field digital mammography (FFDM) systems instead of traditional X-ray screen-film mammography. Digital mammography converts the X rays to digital signals through a solid-state detector and immediately produces the image on a computer monitor. Digital mammograms are easy to transmit, retrieve, and store in a database; in addition, digital images can be manipulated though magnification of a region of interest (ROI), change of contrast, and adjustment of brightness.

In a mammogram, the anatomical features of the breast are superimposed; compression of the breast during the imaging process reduces the superposition of mammary structures by spreading them, and assists in the determination of optimal radiation parameters. The presence of breast cancer is manifested as various signs of abnormality in mammograms; some important signs of breast cancer are described in the following subsections.

1.2.1   MASSES

Breast cancer causes a desmoplastic reaction in the breast tissue [8] and the result is observed in a mammogram as a bright (or dense) object known as a mass. According to the Breast Imaging–Reporting and Data System (BI-RADS®) [13], the shape of masses may be classified as round, oval, lobular, or irregular; the margin of a mass could be circumscribed (well-defined or sharply defined), microlobulated, obscured (hidden by superimposed or adjacent normal tissues and structures), indistinct (ill-defined), or spiculated (lines or spicules appear to be radiating from the margins of the mass). According to their density in comparison with surrounding breast tissue, masses can be described as high density, equal density (isodense), low density (lower attenuation, but not fat-containing), or fat-containing (radiolucent) regions [13].

Based on its morphological features, a dense region (density) or mass may represent a localized sign of breast cancer in a mammogram. A suspicious region in a mammogram is generally categorized as either benign or malignant: a benign mass usually has a well-defined boundary and is round or oval in shape, whereas a malignant tumor has a fuzzy boundary and is irregular in shape [8]. Figure 1.2 shows a mammogram with a benign mass and a mammogram with a malignant tumor; the images are from the mini-Mammographic Image Analysis Society (MIAS) database [14].

1.2.2   CALCIFICATIONS

Calcifications are deposits of calcium in breast tissue and are the most important signs of breast diseases on mammograms [13]. Calcifications appear in mammograms as bright objects and are

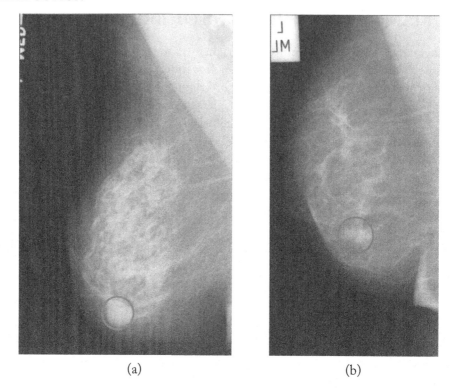

(a)   (b)

Figure 1.2: Examples of mammograms with masses: (a) benign (mdb021); (b) malignant (mdb181). The regions with masses are marked using circles in red. The images are from the mini-Mammographic Image Analysis Society (MIAS) database [14].

generally brighter than masses. Calcifications usually occur in clusters and can be either benign or malignant. Microcalcifications are categorized as more likely to be malignant or benign based on their spatial arrangement in clusters, the amount or number of microcalcifications in a specified area, and the changes in the pattern of the microcalcification clusters as compared to previous mammograms of the same breast. A single cluster of round or oval calcifications is typically of the benign type [11]. Malignant calcifications are typically of irregular, pleomorphic, branching, or rod shape, and form multiple clusters. Figure 1.3 shows a mammogram with benign calcifications and a mammogram with malignant calcifications; the images are from the mini-MIAS database [14].

The possible appearances of calcifications are described in the BI-RADS® lexicon [13] as follows: "Benign calcifications are usually larger than calcifications associated with malignancy. They are much coarser, often round with smooth margins and are much more easily seen. Calcifications associated with malignancy are usually very small and often require the use of a magnifying glass to see them well."

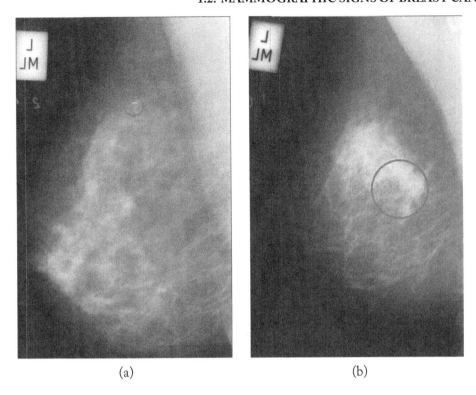

(a)   (b)

Figure 1.3: Examples of mammograms with calcifications: (a) benign (mdb219); (b) malignant (mdb209). The regions with calcifications are marked using circles in red. The images are from the MIAS database [14].

1.2.3 BILATERAL ASYMMETRY

Bilateral asymmetry is the lack of symmetry between images of the left and right breasts of a woman [12]. The appearance of asymmetry may be indicative of a developing mass or merely poor technique during image acquisition. Bilateral asymmetry may cause a modification of the overall appearance of a breast in the mammographic image, illustrated by differences in the distribution of density or the organization of fibrograndular tissue between the left and right breasts, even when pronounced signs of cancer, such as calcifications and masses, are not present. Figure 1.4 shows an example of bilateral asymmetry; the images are from the mini-MIAS database [14].

1.2.4 ARCHITECTURAL DISTORTION

Architectural distortion is a localized sign of breast cancer produced by a desmoplastic reaction [8], manifested by the presence of a distortion in the normal architecture of the breast with no mass visible. Figure 1.5 shows two mammograms with benign and malignant architectural distortion; the

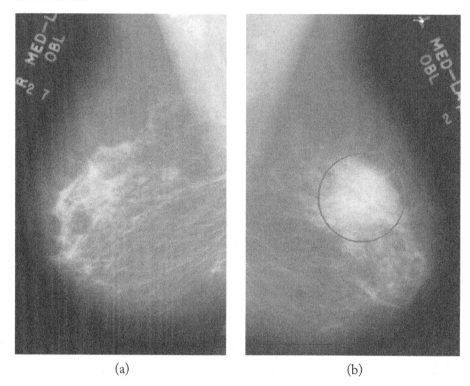

(a)   (b)

Figure 1.4: An example of bilateral asymmetry: (a) right breast (mdb082); (b) left breast (mdb081). The asymmetric region (benign) in the left breast is shown using a circle in red. The images are from the MIAS database [14].

regions of architectural distortion are marked with circles in red. The images are from the mini-MIAS database [14]. Architectural distortion is the third most common sign of nonpalpable breast cancer [15], and a common cause of FN findings on screening mammograms due to its subtle appearance and variability in presentation [16]. See Section 2.1 for further details on architectural distortion.

The presence of an abnormal condition in the breast could be a sign of cancer, precancerous cell formation, or benign breast diseases. Consequently, the major goal during screening or diagnosis is to identify and classify such abnormal regions.

1.3   EVENT DETECTION IN MEDICAL IMAGES

Concepts related to the analysis of the performance of systems for the detection of abnormalities in mammographic images are described in the following subsections.

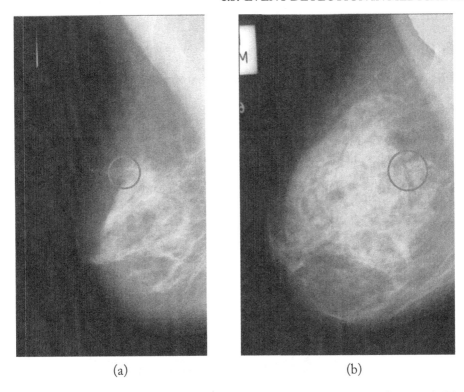

(a)   (b)

Figure 1.5: Examples of mammograms with architectural distortion: (a) benign (mdb127); (b) malignant (mdb125). The regions with architectural distortion are marked using circles in red. The images are from the MIAS database [14].

1.3.1   SENSITIVITY AND SPECIFICITY

Consider a set of ROIs obtained from mammographic images, with some ROIs exhibiting normal tissue patterns (i.e., normal ROIs) and others with some signs of abnormality (i.e., abnormal ROIs). The true type of each mammographic ROI is commonly referred to as the "ground truth," and is correctly established through expert opinion or biopsy. Normal ROIs are labeled as "negative findings" or "negatives." On the contrary, abnormal ROIs are referred to as "positive findings" or "positives."

Consider a computer-aided diagnosis (CAD) system developed to classify positive and negative ROIs. The performance of a CAD system is represented by its sensitivity and specificity. For a system or classifier, sensitivity is defined as the fraction of positives that are identified as positives by the system, and specificity is the fraction of negatives identified as negatives by the system.

For mathematical representation, the following quantities are defined:

- P: the number of positives;

- N: the number of negatives;

- TP: the number of true positives (positives that were categorized as such by the system);

- TN: the number of true negatives (negatives that were categorized as such by the system);

- FP: the number of false positives (negatives that were categorized as positives by the system); and

- FN: the number of false negatives (positives that were categorized as negatives by the system).

An illustration of the relationship between the various quantities is shown in Table 1.1. From the classification matrix of Table 1.1, the following parameters can be defined for the analysis of performance of a CAD system.

Table 1.1: Relationship between the ground truth and the classification results in an experiment. P: positives. N: negatives. TP: true positives. TN: true negatives. FP: false positives. FN: false negatives.

		Ground truth	
		positive	negative
System outcome	positive	TP	FP
	negative	FN	TN
Total		P	N

Sensitivity or True-positive Ratio (TPR):

$$\text{Sensitivity} = \frac{TP}{P} = \frac{TP}{TP + FN}.$$

Sensitivity or TPR indicates how sensitive a system is in the detection of abnormal events; a highly sensitive system will rarely miss an abnormal event. A high sensitivity is the most desirable property of a CAD system.

Specificity or True-negative Ratio (TNR):

$$\text{Specificity} = \frac{TN}{N} = \frac{TN}{TN + FP}.$$

Specificity or TNR indicates how specific the system is in the detection of normal events; a highly specific system will rarely classify a normal event as abnormal and will result in a low rate of false alarms.

False-positive Ratio (FPR):

$$FPR = \frac{FP}{N} = \frac{FP}{TN + FP} = 1 - \text{specificity}.$$

FPR represents the rate of false alarm; a good CAD system should produce minimal or no false alarms. In a breast cancer screening program, false alarms usually lead to increased number of medical tests and biopsies, unnecessary recalls, anxiety on part of the subject, and additional costs to the health-care provider or the patient.

False-negative Ratio (FNR):

$$FNR = \frac{FN}{P} = \frac{FN}{TP + FN} = 1 - \text{sensitivity}.$$

This parameter indicates the rate of missed cases of cancer in a screening program. Ideally, a CAD system should not miss any case of cancer. An FN detection could delay the treatment and may lead to the patient's death. Delayed detection of cancer could make the treatment expensive with reduced chance of survival for the patient.

A good CAD system should simultaneously have a high sensitivity (rarely miss an abnormal event when it occurs) and a high specificity (a low false-alarm rate). Ideally, a CAD system should achieve sensitivity and specificity values of 100%.

Another measure of accuracy of a diagnostic system is the positive-predictive value (PPV), defined as

$$PPV = \frac{TP}{TP + FP}.$$

This measure indicates the probability that a case or ROI identified as positive (abnormal) by a CAD system is, in fact, positive as confirmed by biopsy.

1.3.2 RECEIVER OPERATING CHARACTERISTICS

The receiver operating characteristic (ROC) curve is a graph of all possible values of sensitivity and specificity that are obtained by varying a threshold value that determines the diagnosis [17, 18]. Typically, an ROC curve is plotted within a unit square, with the sensitivity values on the ordinate (y-axis), and $FPR = 1 - \text{specificity}$ on the abscissa (x-axis). The area under the ROC curve (AUC) is denoted by A_z, and is a measure of the overall performance of a system. A perfect classifier results in $A_z = 1$; on the contrary, a random classifier (i.e., each element is randomly labeled by the classifier) produces $A_z = 0.5$. ROC analysis [17, 18] is widely used to evaluate the performance of a radiologist or a CAD system in the detection of abnormalities in medical images. A detailed discussion on ROC analysis is presented in Section 6.2.2.

1.3.3   FREE-RESPONSE RECEIVER OPERATING CHARACTERISTICS

A detection system may provide a discriminant value for each of the detected marks, and a threshold value may be used to determine the acceptance or rejection of such a mark. Change of the threshold value will affect the sensitivity as well as the number of FPs per image. In this context, the free-response ROC curve, abbreviated as the FROC curve, can be used to illustrate the relationship between the sensitivity and the number of FPs per image [19–21].

In FROC analysis, the observer produces binary responses: TP responses (abnormal events detected within an acceptable degree of uncertainty) and FP responses [21]; the FROC data are displayed on a plot with the sensitivity on the ordinate and the average number of FP responses per image or set of images of a patient on the abscissa. The ordinate of the FROC curve is limited to the range of $[0, 1]$ and the abscissa is an open axis of all positive values. FROC analysis can be applied if there is no specific number of TNs [21]. More details on FROC analysis are presented in Section 6.2.3.

1.4   COMPUTER-AIDED DIAGNOSIS OF BREAST CANCER

Although mammography is, at present, the best available tool for early detection of breast cancer, interpreting screening mammograms is a difficult problem resulting in limited sensitivity of screening mammography [22]. The quality of screening mammography is based on measures such as sensitivity, specificity, FPR, recall rate (the number of screening cases called back for further tests), and PPV. The sensitivity of screening mammography is affected by image quality, the volume or number of cases examined in a limited amount of time, and the radiologist's level of expertise; the lack of a number of expert radiologists to analyze mammograms in remote areas is another matter of concern [23]. The estimated sensitivity of screening mammography has been found to vary between 85% and 90% [22]; misinterpretation of breast cancer signs corresponds to 52% of the errors and overlooking signs accounts for 43% of the missed abnormalities.

Minimal signs of abnormalities have been observed to be present on screening mammograms taken previously in some cases of screen-detected cancers [24]. In a study conducted by Blanks et al. [25], double reading of screening mammograms provided greater sensitivity than single reading without increasing recall rates; however, the number of expert radiologists required to perform double reading in a health-care system makes such an approach unrealizable. CAD techniques [2, 26–28] could offer an effective alternative to double reading as a means of reducing errors.

CAD techniques could provide a "second opinion" to the radiologist and help in increasing the sensitivity and accuracy of detection [26, 27]. Various CAD techniques and systems could be as effective as double reading [26, 27] and have been shown to be efficient in detecting frequently observed signs of breast cancer such as masses and calcifications [29]. A CAD system could act as a second reader, and suggest to the radiologist to review areas in a mammogram found to be suspicious by specialized algorithms. A CAD session usually works as follows [23, 27]:

1. The radiologist analyzes the given mammogram and records any questionable or suspicious observations. The radiologist could digitally enhance the given mammogram, if required, in order to give a closer look to subtle details.

2. The CAD system processes the mammogram to detect any signs of abnormality and marks the locations of suspicious regions, if detected.

3. The radiologist then looks at the results provided by the CAD system, and analyzes the areas marked by the system (if any) to verify whether any suspicious region or sign was missed in the first reading.

In addition, CAD systems may be employed to estimate the likelihood that a detected lesion is malignant or benign, or to provide a confidence score through quantitative characterization of the lesion; such an estimate or score is reviewed subsequently by the radiologist.

The potential benefits of CAD-assisted reading motivated the development of several commercial CAD systems, such as the "ImageChecker" (R2 ImageChecker Digital CAD, Hologic, Bedford, MA [30]) and "SecondLook" (iCAD, Nashua, NH [31]); they are being evaluated for their benefits in a screening or diagnostic environment. Studies have shown that CAD systems can improve a radiologist's sensitivity without a substantial increase in the recall rate [32].

Several studies have been conducted to evaluate the effectiveness of the use of CAD systems in a screening program. Ciatto et al. [33] performed comparisons between the outcomes of conventional reading of mammograms and CAD-assisted reading in a national proficiency test of screening mammography in Italy. The performance of single reading with CAD was found to be similar to that of double reading.

Freer and Ulissey [34] conducted a prospective study on the effectiveness of CAD in a screening program; 12,860 screening mammograms were interpreted with the help of a CAD system over a 12-month period. With CAD-assisted reading, the number of cancers detected was increased by 19.5%, the proportion of early-stage malignancies detected was increased from 73% to 78%, the recall rate was increased from 6.5% to 7.7%, and the PPV of biopsy remained unchanged at 38%. The results suggested that CAD can be used to improve the detection of early-stage malignancies without any adverse effect on the recall rate or the PPV of biopsy.

The much debated limitation of CAD for breast cancer is the large number of FPs that CAD systems produce, causing a number of unnecessary recalls and biopsies of healthy participants in the screening program. The recall process results in a traumatic experience and financial strain for the patients and increases the cost to the health-care system. Research has produced inconsistent findings on CAD's sensitivity and effects on recall rate. A study by Fenton et al. [35] indicated that the use of CAD technology resulted in lower specificity, reducing the overall accuracy of interpretation of screening mammograms, while increasing the rate of biopsy.

However, the extent of errors due to missed signs of abnormality strengthens the need for the development of CAD techniques in mammography [2, 26–28]. Various CAD techniques and systems have been developed to improve the sensitivity of detection of breast cancer. Although several CAD techniques are effective in detecting masses and calcifications, they have demonstrated poor performance in the detection of architectural distortion [36]. Increasing the sensitivity and accuracy in the detection of architectural distortion could lead to the desired improvement in the prognosis of breast cancer patients [37] and help in reducing the associated mortality rate.

1.5 ANALYSIS OF PRIOR MAMMOGRAMS

In the context of a screening program, a "detection mammogram" refers to a mammogram on which cancer is detected, and the term "prior mammogram" refers to a mammogram acquired at the last scheduled visit to the screening program prior to the detection of cancer [38]. When breast cancer is detected in a screening program in a particular individual, the case is referred to as "screen-detected cancer." The term "interval cancer" indicates a case where breast cancer is detected outside the screening program in the interval between scheduled screening sessions.

The sensitivity of screening mammography is limited [22]. Minimal signs of abnormality have been found on some prior mammograms of screen-detected cancers [24]. Such cases of abnormality include subtle or hard-to-detect features or patterns that can indicate signs of early breast cancer. Only a few studies have been reported on the analysis of prior mammograms to explore the possibilities of detection of signs of early breast cancer [24, 38–43]. Simultaneous analysis of current and prior mammograms is recommended for use by radiologists in the detection of breast cancer [44–46]. See Section 2.3 for a detailed review on this subject.

Based on the hypothesis that screening mammograms obtained prior to the detection of breast cancer could contain subtle signs of early stages of breast cancer, studies on prior mammograms of screen-detected or interval-cancer cases [38–40, 42, 43] could help in developing strategies for the detection and treatment of breast diseases at their early stages and lead to improvement in the prognosis [37].

1.6 ORGANIZATION OF THE BOOK

Architectural distortion is a subtle abnormality: a sign of breast cancer most commonly missed by radiologists in FN cases of screening programs [36]. In the present work, new features and techniques are developed for the characterization and detection of architectural distortion in prior mammograms of interval-cancer cases.

1.6.1 HYPOTHESIS AND AIM OF THE WORK

CAD systems offer the possibility of increased sensitivity in the detection of breast cancer [2, 26–28]. A CAD system acts as a second reader, calling the attention of the radiologist to areas in mammograms deemed to be suspicious by CAD algorithms. In spite of limited success in the detection of subtle abnormalities, the development of new algorithms for CAD of breast cancer is an active research field [2, 28]. A substantial record of research can be found in the literature on the detection and classification of masses and calcifications: the problems are generally considered to be well investigated and well understood. Any new techniques developed for the detection of masses or calcifications are expected to meet or exceed the high standards of performance set by the existing algorithms. In addition, several commercial CAD systems have demonstrated the ability to detect masses and calcifications with a satisfactory degree of effectiveness. Nevertheless, certain areas of CAD of breast cancer still demand attention; the available CAD systems and methods have been found to fail in the detection of elusive signs of breast cancer such as architectural distortion [38].

As compared to the number of records that exist in the literature regarding studies conducted on the detection of masses and calcifications, a relatively insignificant number of studies have been reported on the problem of detection of architectural distortion in the absence of a central mass. Most of the techniques developed to date are directed toward a more general category of abnormalities, such as stellate lesions that include some of the possible appearances of architectural distortion. More attention is required on the analysis of bilateral asymmetry and curvilinear structures (CLS) for further improvement in the detection of subtle abnormalities [23, 28]. In a larger context, the development of systems for content-based retrieval of mammograms [47–49], indexed atlases [50, 51], and data-mining systems [52] could also be included in the areas of interest related to CAD of breast cancer. FFDM and digital breast tomosynthesis systems could facilitate further studies and development of applications of the techniques mentioned above.

The survival rate of breast cancer patients can be improved significantly via mammographic screening programs. Mammographic screening programs have been proven to be effective in reducing breast cancer mortality rates through early detection of signs of breast cancer. In a screening program, a radiologist detects architectural distortion by identifying faint or subtle signs of abnormality, such as the presence of spiculations and distortion of the normal oriented textural pattern in a mammogram. An increase in the sensitivity of detection of early stages of breast cancer (without affecting the specificity) can be achieved through double reading; but the manpower required renders this approach impractical. Consequently, development of effective CAD systems is the only alternative to facilitate the same level of accuracy.

The hypothesis underlying the present work is that *screening mammograms obtained prior to the detection of breast cancer could contain subtle signs of early stages of breast cancer, in particular, architectural distortion.* In this work, methods are presented for the detection of sites of architectural distortion in prior mammograms of interval-cancer cases in a screening program using several image processing and pattern classification methods. Simultaneous analysis of the current and prior mammograms is usually recommended by radiologists, and could be helpful in the detection of early signs of breast cancer by radiologists; the same approach could also help to enhance the performance of CAD systems [44, 46]. The development of CAD systems designed for the detection of subtle or hard-to-detect signs, in particular, architectural distortion in prior mammograms, could improve the prognosis of breast cancer patients by facilitating the detection of breast diseases at their early stages.

Improvement in the detection rate of architectural distortion could increase the rate of detection of early breast cancer, and reduce the morbidity and mortality due to the disease [53].

1.6.2   STRUCTURE OF THE BOOK

This book presents new and original techniques for the extraction and analysis of features for the characterization and detection of architectural distortion in prior mammograms, as well as analysis of various methodologies for CAD of early signs of breast cancer. The book is structured as follows:

Chapter 2 presents a review of the state of the art in CAD of architectural distortion and analysis of prior mammograms.

Chapter 3 provides an introduction to the analysis of oriented patterns using Gabor filters and phase portraits.

Chapter 4 presents techniques for the detection of potential sites of architectural distortion in mammograms using Gabor filters and analysis of the phase portraits of the related oriented patterns.

Chapter 5 gives the details of the experimental set up and datasets used in the present work.

Chapter 6 presents descriptions of the theoretical and practical aspects of several methods for feature selection, feature analysis, cross-validation, and pattern classification.

Chapter 7 presents the results of analysis of features for the characterization and detection of architectural distortion, including the original contributions of this work, such as the geometrical transformation of ROIs; structural analysis of patterns of architectural distortion via Laws' texture energy measures; characterization of the angular spread of power in the frequency domain and also using Gabor magnitude response, angle response, coherence, and orientation strength; and the use of higher order Rényi and Tsallis entropy measures for the characterization of angular distributions.

Chapter 8 provides a detailed discussion of the results, performance analysis, and statistical analysis of various aspects of the detection of architectural distortion in prior mammograms including cross-validation with multiple datasets.

Chapter 9 presents concluding remarks on the work as well as a discussion of future research problems in the detection of architectural distortion in mammograms.

CHAPTER 2

Detection of Early Signs of Breast Cancer

2.1 DETECTION OF ARCHITECTURAL DISTORTION

Architectural distortion (i.e., a distortion of the architecture of breast parenchyma without being accompanied by increased density or a mass) is the third most common mammographic sign of non-palpable breast cancer [15], and is an important finding in interpreting breast cancer [54]. However, due to its subtlety and variability in presentation, this sign of malignancy is often missed during screening [15]. The detection of architectural distortion is performed by a radiologist through the identification and localization of subtle signs of abnormality, including the presence of spiculations and distortion of the normal texture of the breast. CAD techniques and systems have been proven to be effective in detecting masses and microcalcifications with satisfactory performance, but they have been found to fail in detecting architectural distortion with an adequate level of accuracy [36]. Increase of sensitivity and accuracy in the detection of architectural distortion could lead to an effective improvement in the prognosis of breast cancer patients [37], and thus help in reducing the mortality rate due to the disease.

In a mammogram, the breast is seen in the form of an image with oriented texture [12] due to the presence of a variety of normal piecewise linear structures (e.g., vessels, ducts, and fibroglandular tissue) [55]. The normal oriented texture pattern, which typically converges toward the nipple, is changed or distorted in the presence of architectural distortion. Architectural distortion is defined in BI-RADS® [13] as follows: "The normal architecture (of the breast) is distorted with no definite mass visible. This includes spiculations radiating from a point and focal retraction or distortion at the edge of the parenchyma. Architectural distortion can also be an associated finding."

According to Homer [8], "Architectural distortion is a localizing sign of breast cancer produced by a desmoplastic reaction; its presence demands an explanation. Some benign etiologies for this finding, such as previous biopsy and inflammation can be suspected by history. If no explanation for the architectural distortion can be elicited, biopsy is often the next indicated procedure." Focal retraction is usually considered to be easier to recognize than spiculated distortion within the breast parenchyma [8].

Indirect signs of malignancy (such as architectural distortion, bilateral asymmetry, single dilated duct, and developing densities) account for almost 20% of the detected cancers [53]. Because architectural distortion may mimic the appearance of normal breast tissue overlapped in the projected mammographic image, its detection could be difficult and it is the most commonly missed abnormality in FN cases [16]. Specifically, architectural distortion accounts for 12% to 45% of breast cancer cases overlooked or misinterpreted in screening mammography [56, 57].

Architectural distortion could be due to malignant or benign diseases and abnormalities; the malignant class includes cancer, and the benign class includes scar and soft-tissue damage due to trauma. Examples of benign and malignant architectural distortion are shown in Figure 1.5. Architectural distortion could appear at the initial stages of the formation of a breast mass or tumor and has been found to be associated with breast malignancy in one-half to two-thirds of the cases in which it is present [54]; as such, architectural distortion is an important finding in interpreting the manifestation of breast cancer on mammograms [8].

Karssemeijer and te Brake [58] proposed a multiscale-based method for the detection of stellate distortion including spiculating masses and architectural distortion using the output of three-directional, second-order, Gaussian derivative operators with the direction of the filters differing by $\pi/3$ in orientation, and obtained a sensitivity of about 90% at the rate of one FP/image. Mudigonda and Rangayyan [59] proposed the use of texture flow-field based on the local coherence of texture orientation to detect architectural distortion; preliminary results indicated the potential of the proposed method in the detection of architectural distortion.

Ayres and Rangayyan [60–63] and Rangayyan and Ayres [64] proposed methods based on the application of Gabor filters and phase-portrait modeling to characterize subtle changes due to architectural distortion from a pattern recognition perspective. The methods were applied to two datasets, one set with 19 cases of architectural distortion and 41 normal mammograms from the MIAS database [14], and another set with 37 cases of architectural distortion. Sensitivity rates of 84% at 4.5 FP/image and 81% at 10 FP/image were obtained from FROC analysis for the two sets of images [60].

Matsubara et al. [54, 65] used morphological image processing techniques along with a concentration index to detect architectural distortion around the skin line and within the mammary gland; the sensitivity obtained was 94% with 2.3 FP/image and 84% with 2.4 FP/image, respectively. Ichikawa et al. [66] presented an automatic method for the detection of areas related to spiculated architectural distortion; suspicious areas were detected by means of a concentration index of linear structures obtained using the mean curvature of the given image. Discriminant analysis was performed with the nine features obtained for classification, and a sensitivity of 68% with 3.4 FP/image was obtained. Hara et al. [67] used dynamic range compression as a preprocessing step before extracting the mammary gland by a combination of mean curvature and a shape index; a sensitivity of 70% was achieved at 2 FP/image. Matsubara et al. [68] proposed a modification of their previous method for the detection of architectural distortion using the mean curvature of images with a combination of two levels of resolution after dynamic range compression, and improved the accuracy of extraction for thin mammary glands; at the final stage, they obtained a sensitivity of 75% at 2.9 FP/image.

Guo et al. [69] studied the characterization of architectural distortion using the Hausdorff fractal dimension (FD) and performed classification of ROIs (exhibiting architectural distortion and those with normal mammographic patterns) using a support vector machine (SVM). A set of 40 ROIs was selected from the MIAS database [14], including 19 ROIs with architectural distortion and 21 ROIs with normal tissue patterns. The best classification accuracy obtained was 72.5%. Guo et al. [70] also used five different methods to estimate the FD and an SVM to differentiate masses and architectural distortion from normal parenchyma; using FD and lacunarity, the best result obtained for architectural distortion in terms of AUC was 0.875 ± 0.055.

Sampat and Bovik [71] and Sampat et al. [72, 73] applied a linear filter to the Radon transform of the given image for the enhancement of spicules; the enhanced image was filtered with radial spiculation filters to detect spiculated masses and architectural distortion marked by converging lines or spiculation. Using a set of 45 images with spiculated masses and another set of 45 images with architectural distortion, the sensitivity achieved was 91% at 12 FP/image and 80% at 14 FP/image, respectively. Özekes et al. [74] used several distance thresholds to detect architectural distortion and reported an accuracy of 89.02%.

Tourassi et al. [55] investigated the use of FD to distinguish between normal tissue patterns and architectural distortion in mammographic ROIs. The FD was estimated using the circular average power spectrum technique [75, 76]. The method was applied to a dataset of 1,500 ROIs, including 112 ROIs with architectural distortion and 1,388 ROIs exhibiting normal tissue patterns. The best performance achieved was 0.89, in terms of AUC. Tourassi et al. observed that the presence of architectural distortion disrupts the self-similarity properties of breast parenchyma, and that the average FD of the ROIs with architectural distortion was significantly lower than that of normal ROIs.

A method to detect masses and architectural distortion by locating points surrounded by concentric layers of image activity was proposed by Eltonsy et al. [77]. The method was tested on a set of 80 images, including 13 masses, 38 masses with architectural distortion, and 29 images with only architectural distortion. The overall sensitivity reported was 91.3% at 9.1 FP/image; however, in the detection of architectural distortion, a sensitivity of 93.1% was obtained at the same rate of FPs.

Nakayama et al. [78] performed multiresolution analysis by decomposing the original digitized image into several subimages at three scales by a novel filter bank based on wavelets and the Hessian matrix. With six objective features obtained from automatically detected ROIs at three scales, the sensitivity obtained was 71.3% (57 out of 80 images) at 3.01 FP/image.

Nemoto et al. [29] proposed a method to detect architectural distortion with radiating spiculations. The methods are based on the observation that the lines corresponding to spiculation of architectural distortion differ in characteristics from the lines in the normal mammary gland. The likelihood of spiculation was computed and a modified point-convergence index weighted by the likelihood was calculated to enhance architectural distortion. After the classification step, a sensitivity of 80.0% was obtained at 0.80 FP/image.

Jasionowska et al. [79] proposed a method consisting of two stages involving the detection of ROIs with potential architectural distortion based on analysis with Gabor filters and the recognition of architectural distortion using the two-dimensional (2D) Fourier transform in polar coordinates. The method was tested with 33 mammograms containing architectural distortion from the Digital Database for Screening Mammography (DDSM) [80] and a sensitivity of 68% with 0.86 FP/image was obtained.

Table 2.1 presents a few important points regarding some of the studies related to the detection of architectural distortion; only works reporting results with the AUC or A_z and/or analysis of FROC are listed.

2.2   DETECTION OF ARCHITECTURAL DISTORTION BY CAD SYSTEMS

CAD techniques [2, 26–28] could offer a cost-effective alternative to double reading as a means of reducing errors. CAD systems act as a second reader and help in the decision making process by drawing the attention of radiologists to review suspicious areas in a mammogram; however, the final decision is always made by the radiologist.

Only a small number of studies have been reported on analysis of the performance of commercial CAD systems in the detection of architectural distortion [38]. Burhenne et al. [81] studied the performance of a commercial CAD system for mammography and obtained a sensitivity of 75% in the detection of masses and architectural distortion at one FP/image. Evans et al. [82] investigated the ability of a commercial CAD system to mark invasive lobular carcinoma of the breast, and obtained a sensitivity of 91% with screening mammograms demonstrating biopsy-proven cancer, and 77% with the corresponding prior mammograms. Birdwell et al. [83] studied the performance of a commercial CAD system used for marking the signs of cancer that were overlooked or missed by radiologists; the system was able to detect five out of six cases of architectural distortion, and 77% of the previously missed lesions, at 2.9 FP/image.

Baker et al. [36] studied the performance of two commercially available CAD systems in detection of architectural distortion; fewer than half of the 45 cases of architectural distortion were detected, resulting in a lower image-based sensitivity of 38% (or 30 out of 80 images) at 0.7 FP/image. These findings indicate the need for further research in this area, and the development of CAD algorithms designed specifically for the characterization and detection of architectural distortion.

2.3   DETECTION OF SIGNS OF CANCER IN PRIOR MAMMOGRAMS

Screening mammography has limited sensitivity [22]; it has been observed that subtle signs of abnormality can be identified in a significant portion of prior mammograms of cases of screen-detected cancer [38] or interval-cancer cases [24, 40–43, 84–87]. Such cases of abnormality include subtle or hard-to-detect features or patterns that can indicate early signs of breast cancer. Only a few studies have been reported on the analysis of prior mammograms to explore the possibilities of detection of early signs of breast cancer [24, 38–41, 84–87].

Table 2.1: Performance statistics of selected methods for the detection of architectural distortion. Only studies with ROC analysis, FROC analysis, and/or quantitative results of pattern classification are listed. See also Table 2.2. Reproduced with permission from: S. Banik, R. M. Rangayyan, and J. E. L. Desautels, "Digital Image Processing and Machine Learning Techniques for the Detection of Architectural Distortion in Prior Mammograms." In K. Suzuki, Ed., *Machine Learning in Computer-aided Diagnosis: Medical Imaging Intelligence and Analysis*, pp. 24–63, IGI Global, Hershey, PA, January 2012. © IGI Global.

Authors	Size of Dataset	Summary of Methods and Results
Karssemeijer and te Brake [58]	31 normal cases and 19 cases with stellate lesions from the MIAS database [14].	A multiscale-based method for the detection of stellate distortion including spiculating masses and architectural distortion using the output of three-directional, second-order, Gaussian derivative operators; a sensitivity of about 90% at one FP per image.
Guo et al. [69]	40 ROIs including 19 with architectural distortion and 21 with normal tissue patterns from the MIAS database [14].	Hausdorff FD and an SVM; classification accuracy of 72.5%.
Guo et al. [70]	19 ROIs with architectural distortion and 41 with normal breast parenchyma, manually selected from the MIAS database [14].	Lacunarity and five methods to estimate the FD with an SVM to differentiate masses and architectural distortion from normal parenchyma; $AUC = 0.875 \pm 0.055$.
Sampat et al. [72,73]	45 cases with spiculated masses and 45 cases with architectural distortion.	A linear filter and Radon transform for the enhancement of spicules; sensitivity of 80% at 14 FP/image.

Table 2.1: *Continued.*

Authors	Size of Dataset	Summary of Methods and Results
Tourassi et al. [55]	112 ROIs with architectural distortion and 1,388 normal ROIs.	FD estimated with the circular average power spectrum technique; $A_z = 0.89$.
Matsubara et al. [65]	55 mammograms with architectural distortion (17 with focal retraction, 38 with architectural distortion within the fibroglandular disk).	Morphological image processing with a concentration index. Around the skin line: sensitivity of 94% with 2.3 FP/image; within the mammary gland: sensitivity of 84% with 2.4 FP/image.
Ichikawa et al. [66]	94 mammograms with architectural distortion.	Automatic detection of spiculated architectural distortion using a concentration index of linear structures obtained by the mean curvature of the image. Sensitivity of 68% with 3.4 FP/image.
Hara et al. [67]	99 cases of architectural distortion.	Dynamic range compression before extracting the mammary gland by a combination of mean curvature and a shape index; sensitivity of 70% at 2 FP/image.
Matsubara et al. [68]	280 mammograms (121 abnormal mammograms with architectural distortion and 159 normals) from the DDSM [80].	A modification of previous methods for the detection of architectural distortion in thin mammary glands; sensitivity of 75% at 2.9 FP/image.
Eltonsy et al. [77]	80 images (13 masses, 38 masses accompanied by architectural distortion, and 29 with only architectural distortion).	A method for locating points surrounded by concentric layers of image activity; overall sensitivity of 91.3% with 9.1 FP/image; sensitivity of 93.1% for architectural distortion.

Table 2.1: *Continued.*

Authors	Size of Dataset	Summary of Methods and Results
Özekes et al. [74]	Cases of architectural distortion from the mini-MIAS [14] database.	Several distance thresholds; an accuracy of 89.02%.
Ayres and Rangayyan [63]	106 ROIs (17 cases of architectural distortion, 45 normals, two ROIs with malignant calcifications, and 44 masses).	A bank of Gabor filters and linear phase-portrait modeling to characterize subtle changes due to architectural distortion; sensitivity 76.5%, specificity 76.4%, and $A_z = 0.77$.
Ayres and Rangayyan [60]	19 images containing architectural distortion and 41 normal mammograms.	A bank of Gabor filters and constrained linear phase-portrait modeling; sensitivity of 84% at 4.5 FP/image.
Nakayama et al. [78]	80 mammograms with architectural distortion.	Multiresolution analysis by decomposing the original image into several subimages at three scales by a novel filter bank based on wavelets and the Hessian matrix; sensitivity of 71.3% (57 out of 80 images) at 3.01 FP/image.
Nemoto et al. [29]	25 computed radiography (CR) digital mammograms, each with a single area of radiating spiculation of architectural distortion.	The likelihood of spiculation and a modified point-convergence index weighted by the likelihood to enhance architectural distortion; sensitivity of 80.0% at 0.80 FP/image.

Banik et al. [40] proposed methods based upon Gabor filters, phase-portrait analysis [63, 64, 88, 89], analysis of the angular spread of power [84], fractal analysis [38, 75, 76], Laws' texture energy measures [87, 90], and Haralick's texture features [39, 91, 92] for the detection of architectural distortion in prior mammograms of interval-cancer cases, and reported a sensitivity of 0.80 at 5.8 FP/image.

Sameti et al. [93] studied the structural differences between the regions that subsequently developed malignant masses on mammograms and other normal areas in images taken in the last screening instance prior to the detection of tumors. Manually identified circular ROIs were transformed into their optical-density equivalent images, and divided into three types of regions corresponding to low, medium, and high optical density. Based upon the regions, a set of photometric and texture features was extracted. Differences were observed between regions related to malignant tumors and normal tissues in the prior mammograms in 72% of the 58 cases of breast cancer studied. Sameti et al. [41] also reported an average classification rate of 72% using six selected texture and photometric features computed from manually marked regions on the last screening mammograms prior to the detection of breast cancer.

Rangayyan et al. [38] used phase portraits, FD, and Haralick's texture features for the detection of architectural distortion in prior mammograms of screen-detected cancer, and achieved a sensitivity of 79% at 8.4 FP/image with a set of 14 prior mammograms.

Petrick et al. [94] studied the effectiveness of their method for the detection of benign and malignant masses as applied to the related regions in prior mammograms. Using a set of 92 images, including 54 malignant and 38 benign lesions from 37 cases (22 malignant and 15 benign), the methods provided a "by film" mass-detection sensitivity of 51% with 2.3 FP/image. A slightly better accuracy of 57% was achieved in the detection of malignant tumors.

A method for the detection of masses in current and prior mammograms was proposed by Zheng et al. [95]. The method was applied in two situations: the algorithm optimized using the current mammograms, and the algorithm optimized using the related prior mammograms. The method consisted of three steps: initial selection of the potential sites of lesions using difference-of-Gaussian filtering and thresholding, elimination of FPs using adaptive region growing and topological analysis of the suspicious regions, and finally, extraction of shape, histogram, and texture features followed by classification using an artificial neural network (ANN) classifier. The method was tested on a database of 260 pairs of consecutive mammograms, with the latest image containing one or two masses and the prior image originally being classified as negative or probably benign. The first two steps of the method were applied to both the latest and prior mammograms, resulting in the extraction of a set of 1,449 suspicious ROIs. The ROIs were classified based on the true mass location in the corresponding latest mammograms. Training the ANN with the ROIs from the latest mammograms resulted in $AUC = 0.89 \pm 0.01$ and 0.65 ± 0.02 in the classification of ROIs from the latest and prior mammograms, respectively. On the other hand, training the ANN with ROIs from the prior mammograms resulted in $AUC = 0.81 \pm 0.02$ and 0.71 ± 0.02 in the classification of ROIs from the latest and prior mammograms, respectively. The results demonstrate the importance of incorporating knowledge about particular features of early signs in developing CAD algorithms for the detection of early signs of breast cancer.

Burnside et al. [96] evaluated the effect of the availability of prior mammograms on the clinical outcomes of diagnostic and screening mammography, and reported that incorporating prior mammograms improved the specificity of screening mammography significantly but did not improve the sensitivity. However, the approach increased the sensitivity of diagnostic mammography.

Sumkin et al. [44] assessed and compared the benefit of using images acquired one year or two years previously during the interpretation of current mammograms; it was found that the sensitivity was not significantly affected by the availability of the prior mammograms, but the specificity was improved. Varela et al. [45] found that the use of prior mammograms as reference could significantly increase the accuracy of classification between benign and malignant masses. Ciatto et al. [97] compared the outcomes of single, double, and CAD-assisted reading of negative prior mammograms in cases of interval cancer; the study suggested that CAD-assisted reading produces significantly more specific results and is almost as sensitive as double reading. Evans et al. [82] investigated the performance of a commercial CAD system in detecting invasive lobular carcinoma of the breast, and reported a sensitivity of 91% with screening mammograms demonstrating biopsy-proven cancer, and 77% with the corresponding prior mammograms.

Moberg et al. [98] conducted a study on CAD-assisted analysis of cases of interval cancer and reported that although the CAD system had high sensitivity, it had no effect on the sensitivity or the specificity of the radiologists. Ikeda et al. [99] analyzed the performance of a commercial CAD system using prior mammograms of 172 cases of cancer with subtle findings; the system was able to detect 42% of the findings. Garvican and Field [100] conducted a study on the performance of a commercial CAD system with prior mammograms of interval-cancer cases; in the analysis of difficult cases, the system was found to "over-prompt" normal areas and "under-prompt" the areas with cancer.

Table 2.2 summarizes some of the reported studies on prior mammograms; only studies reporting results with the ROC and/or FROC analysis are listed.

Simultaneous analysis of the current and prior mammograms is recommended for use by radiologists in the detection of breast cancer [44–46]; the same approach could be used improve the performance of CAD systems [38]. Complementary to systems designed for the detection of well-developed masses or calcifications, the development of CAD systems designed specifically for the detection of subtle or hard-to-detect signs in prior mammograms, in particular, architectural distortion, is important and could improve the prognosis by facilitating the detection of breast diseases at their early stages.

2.4 REMARKS

Screening and interpretation of mammographic images is a difficult task; the large variability in the appearance and subtle nature of abnormalities, occlusion of abnormalities in dense breast tissues, and the need for making accurate decisions impose additional challenges on the radiologists.

Table 2.2: Performance statistics of selected methods for the analysis of prior mammograms. Only studies with ROC analysis, FROC analysis, and/or pattern classification results are listed. Reproduced with permission from: S. Banik, R. M. Rangayyan, and J. E. L. Desautels. "Digital Image Processing and Machine Learning Techniques for the Detection of Architectural Distortion in Prior Mammograms." In K. Suzuki, Ed., *Machine Learning in Computer–aided Diagnosis: Medical Imaging Intelligence and Analysis*, pp. 24–63, IGI Global, Hershey, PA, January 2012. © IGI Global.

Authors	Size of Dataset	Summary of Methods and Results
Rangayyan et al. [38]	A set of 14 prior mammograms with 14 detection mammograms.	Gabor filters, phase portraits, FD, and texture features for the detection of architectural distortion in prior mammograms of screen-detected cancer; sensitivity of 79% at 8.4 FP/image.
Sameti et al. [41]	Mammograms of 58 biopsy-proven breast cancer patients taken 10 to 18 months prior to cancer detection.	Six selected texture and photometric features from manually marked regions on the last screening mammograms prior to the detection of breast cancer; average classification rate of 72%.
Zheng et al. [95]	260 pairs of consecutive mammograms where the latest image showed one or two masses, and the prior image had been originally classified as negative or probably benign.	Training the ANN with the latest mammograms: $AUC = 0.89 \pm 0.01$ and 0.65 ± 0.02 using the latest and prior mammograms, respectively. Training the ANN with ROIs from the prior mammograms: $AUC = 0.81 \pm 0.02$ and 0.71 ± 0.02 using ROIs from the latest and prior mammograms, respectively.

Table 2.2: *Continued.*

Authors	Size of Dataset	Summary of Methods and Results
Petrick et al. [94]	A set of 92 images, including 54 malignant and 38 benign lesions from 37 cases (22 malignant and 15 benign).	Detection of benign and malignant masses in the related regions in prior mammograms; "by film" mass-detection sensitivity of 51% with 2.3 FP/image; a slightly better accuracy of 57% in the detection of malignant tumors.
Rangayyan et al. [39]	106 prior images of architectural distortion from 56 interval-cancer cases and 52 normal mammograms from 13 individuals.	Gabor filters, phase-portrait analysis, FD, and Haralick's texture features; sensitivity of 0.80 at 7.6 FP/image and $AUC = 0.77$.
Banik et al. [40]	106 prior images of architectural distortion from 56 interval-cancer cases and 52 normal mammograms from 13 individuals.	Gabor filters, phase-portrait analysis, angular spread of power in the frequency domain, FD, Laws' texture energy measures, and Haralick's texture features; sensitivity of 0.80 at 5.8 FP/image and $AUC = 0.78$.

The development of efficient and accurate CAD systems could help the radiologist in the decision making process. The development of CAD techniques for breast cancer is an active research field: several efficient CAD techniques have been developed for the detection of masses and calcifications; however, a significantly smaller numbers of methods have been proposed for the detection of architectural distortion.

This chapter presented a comprehensive review of the literature on CAD techniques for the detection of architectural distortion; in addition, some of the studies on prior mammograms and interval-cancer cases were described. The studies reviewed indicate the need and scope for further developments in the detection of subtle signs of early stages of breast cancer.

CHAPTER 3

Detection and Analysis of Oriented Patterns

3.1 ORIENTED TEXTURE IN BIOMEDICAL IMAGES

Texture is an important characteristic feature of medical images, and the presence of particular types of texture may convey important information about the scene or the objects contained [12]. Oriented texture is commonly encountered in medical images. In this context, the analysis of texture, in particular analysis of oriented texture or patterns, is an important task in the framework of understanding and interpretation of medical images.

The detection of oriented patterns largely depends on the characteristic features of the objects under investigation, the appearance of neighboring structures, superposition of multiple structures, and the presence of noise. Individual oriented features are associated with a particular spatial width, length, orientation, and/or intensity, such as the length or width of a spicule in a mammographic image [23]. Therefore, the ability and accuracy of a method developed for the detection of oriented features relies on the proper calibration or selection of its parameters in relation to the internal characteristics of the oriented features of interest. In addition, noise and other neighboring or ambiguous structures affect the performance of an oriented feature detector; they may lead to the detection of different or irrelevant oriented structures in the image [101].

A number of methods have been developed for the analysis of oriented patterns and detection of lines, including Fourier-transform-based methods [102–105]; space-domain linear filters, such as steerable filters [106], Gabor filters [64, 107–110], and infinite-impulse-response recursive fan filters [111]; and nonlinear filters [112, 113]. This chapter presents the theory and application of real Gabor filters and linear phase-portrait analysis for the detection of CLS and architectural distortion in mammograms [64, 101]. Some of the methods mentioned above were developed by Ayres et al. [23] and have been used in the present work to obtain the potential sites of architectural distortion.

3.2 GABOR FILTERS

Gabor filters consist of a category of filters obtained by the modulation of a Gaussian envelope by a sinusoidal function (real or complex) [107]. The Gabor filter provides the best compromise between spatial localization and frequency localization, as represented by the product between the spatial extent and the frequency bandwidth of the filter [12, 107–109]. The Gabor filter is considered to be a useful tool in computer vision and in signal and image processing. In image processing applications, Gabor filters may be used as line detectors or detectors of oriented features [12, 64, 101, 109, 110].

3.2.1 THE REAL GABOR FILTER

The real Gabor filter kernel oriented at the angle $\theta = -\pi/2$ is given as [64, 101]

$$g(x, y) = \frac{1}{2\pi\sigma_x\sigma_y} \exp\left[-\frac{1}{2}\left(\frac{x^2}{\sigma_x^2} + \frac{y^2}{\sigma_y^2}\right)\right]\cos(2\pi f_o x), \tag{3.1}$$

where σ_x and σ_y are the standard deviation values in the x and y directions, and f_o is the frequency of the modulating sinusoid. Kernels at other angles can be obtained by rotating this kernel over the range $[-\pi/2, \pi/2]$ by using coordinate transformation as

$$\begin{bmatrix} x' \\ y' \end{bmatrix} = \begin{bmatrix} \cos\alpha & \sin\alpha \\ -\sin\alpha & \cos\alpha \end{bmatrix}\begin{bmatrix} x \\ y \end{bmatrix},$$

where (x', y') is the set of coordinates rotated by the angle α. In the present work, a set of 180 kernels was used, with angles spaced evenly over the range $\theta = [-\pi/2, \pi/2]$. The parameters in Equation 3.1 were derived by taking into account the size of the lines or CLS to be detected, as follows [64]:

- Let τ be the full-width at half-maximum of the Gaussian term in Equation 3.1 along the x axis. Then, $\sigma_x = \tau/(2\sqrt{2\ln 2}) = \tau/2.35$.

- Let the period of the cosine term be τ; then, $f_0 = 1/\tau$.

- The value of σ_y is defined as $\sigma_y = l\sigma_x$, where l determines the elongation of the Gabor filter in the y direction, as compared to the extent of the filter in the x direction.

- The parameter τ controls the scale of the filter. In the present work, for the detection of architectural distortion, $\tau = 4$ pixels (corresponding to a thickness of 0.8 mm at the pixel size of 200 μm) and $l = 8$ were used. These values were determined empirically, by observing the typical spicule width and length in mammograms with architectural distortion [64].

Figure 3.1 demonstrates the effects of the different design parameters of the real Gabor filter and illustrates the scaling and rotation phenomena. The design criteria shown in Figure 3.1 are described as follows:

- Figures 3.1 (a) and 3.1 (e) show the impulse response of an initial Gabor filter and its Fourier transform magnitude (frequency response), respectively.

- The Gabor filter shown in Figure 3.1 (b) is obtained by increasing the parameter τ of the original Gabor filter; thus, the impulse response of the filter is thickened. Correspondingly, the Fourier spectrum of the enlarged filter, shown in Figure 3.1 (f), is moved to a lower frequency range along the horizontal frequency axis.

- In Figure 3.1 (c), the Gabor filter of Figure 3.1 (a) is stretched in the vertical direction, by increasing the elongation factor l. The Fourier spectrum of the modified Gabor filter, shown in Figure 3.1 (g), is compressed in the vertical frequency variable to a narrower bandwidth than the original.

- The effect of rotating the original Gabor filter by 45° clockwise is displayed in Figures 3.1 (d) and 3.1 (h), showing the rotated impulse response and its corresponding Fourier spectrum, respectively.

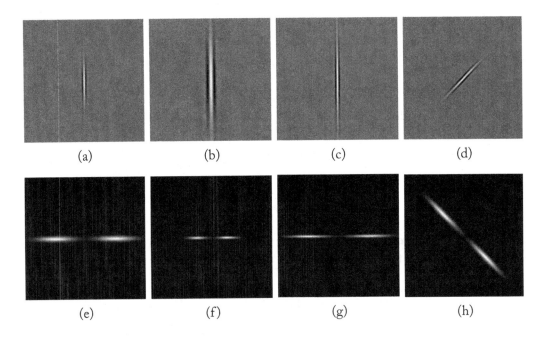

(a)　　　　　(b)　　　　　(c)　　　　　(d)

(e)　　　　　(f)　　　　　(g)　　　　　(h)

Figure 3.1: Effects of the different parameters of the Gabor filter. (a) Example of the impulse response of a Gabor filter. (b) The parameter τ is increased: the thickness of Gabor filter is increased. (c) The parameter l is increased: the Gabor filter is elongated. (d) The angle of the Gabor filter is modified. Figures (e)–(h) correspond to the magnitude of the Fourier transforms of the Gabor filters in (a)–(d), respectively. The $(0, 0)$ frequency component is at the center of the spectra displayed. Reproduced with permission from: S. Banik, R. M. Rangayyan, and J. E. L. Desautels. "Image Processing and Pattern Classification Techniques for the Detection of Architectural Distortion in Prior Mammograms of Interval-cancer Cases." In J. S. Suri, S. V. Sree, K.-H. Ng, and R. M. Rangayyan, Eds., *Diagnostic and Therapeutic Applications of Breast Imaging*, pp. 197–242, SPIE Press, Bellingham, WA, February 2012. © SPIE.

Let $\theta^r(x, y)$ denote the angle of the oriented feature at (x, y), and $g_k^r(x, y)$, $k = 0, 1, \cdots, (K - 1)$, denote the real Gabor filter oriented at $\alpha_k = -\pi/2 + \pi k/K$, where K is the number of equally spaced filters over the angular range $[-\pi/2, \pi/2]$; in this work, $K = 180$ was used. The orientation field can be extracted using the bank of real Gabor filters as represented by the function, $g_k^r(x, y), k = 0, 1, \cdots, (K - 1)$. The filtering operation was designed to be performed in the frequency domain.

Let $I(x, y)$ denote the image to be processed, and $W_k^r(x, y)$ represent the Gabor-filtered images. The orientation field of $I(x, y)$ obtained by the bank of real Gabor filters can be represented by the angle

$$\theta^r(x, y) = \alpha_{k_{\max}} \quad \text{where} \quad k_{\max} = \arg\{\max_k[W_k^r(x, y)]\} ,$$

and by the amplitude of the output of the real Gabor filter at the optimal orientation $M^r(x, y) = W_{k_{\max}}^r(x, y)$. The real Gabor filter, described as above, can be used for the detection of oriented features with positive contrast only.

3.2.2   THE COMPLEX GABOR FILTER

The complex Gabor filter kernel oriented at the angle $\theta = -\pi/2$ is given by

$$
\begin{aligned}
g^c(x, y) &= \frac{1}{2\pi\sigma_x\sigma_y} \exp\left[-\frac{1}{2}\left(\frac{x^2}{\sigma_x^2} + \frac{y^2}{\sigma_y^2}\right)\right] \exp(j2\pi f_0 x) \\
&= g^r(x, y) + jg^i(x, y),
\end{aligned}
\tag{3.2}
$$

where $g^r(x, y)$ is the real Gabor filter as in Equation 3.1, and $g^i(x, y)$ is the imaginary component of the complex Gabor filter, given by

$$g^i(x, y) = \frac{1}{2\pi\sigma_x\sigma_y} \exp\left[-\frac{1}{2}\left(\frac{x^2}{\sigma_x^2} + \frac{y^2}{\sigma_y^2}\right)\right] \sin(2\pi f_0 x) .$$

Similar to the real Gabor filter, complex Gabor kernels at other angles can be obtained by rotating the kernel $g^c(x, y)$ over the range $[-\pi/2, \pi/2]$. The same design rules as described on page 28 apply for the selection of the parameters $\sigma_x, \sigma_y,$ and f_0.

The imaginary part of the complex Gabor filter is an approximation of the Hilbert transform of the real Gabor filter; consequently, the complex Gabor filter acts as a quadrature pair filter. Let $W_k^r(x, y)$ represent the result of filtering a given image $I(x, y)$ with the real Gabor filter $g^r(x, y)$; let $W_k^i(x, y)$ denote the result of filtering $I(x, y)$ with the imaginary component $g^i(x, y)$ of the complex Gabor filter; and let $W_k^c(x, y)$ denote the result of filtering $I(x, y)$ with the complex Gabor filter $g^c(x, y)$.

Then, from Equation 3.2, $|W_k^c(x, y)|^2 = |W_k^r(x, y)|^2 + |W_k^i(x, y)|^2$. The orientation field computed using the complex Gabor filter bank can be given by the angle field $\theta^c(x, y)$, with

$$\theta^c(x, y) = \alpha_{k_{\max}} \quad \text{where} \quad k_{\max} = \arg\{\max_k[|W_k^c(x, y)|]\} ,$$

and by the magnitude of the output of the complex Gabor filter at the optimal orientation $M^c(x, y) = |W^c_{k_{max}}(x, y)|$.

From an analysis of the frequency response, it can be observed that Gabor filters (real and complex) are narrowband filters [64, 101]; the complex Gabor filter detects oriented features of both positive and negative contrast. However, in the present work, only a bank of real Gabor filters was used to detect features with positive contrast; the filtering operation was implemented in the frequency domain.

3.3   ANALYSIS OF ORIENTED PATTERNS USING PHASE PORTRAITS

The phase-portrait diagram of a system of two linear, first-order, differential equations illustrates the possible trajectories of the state variables for different initialization values [88]. Rao and Jain [89] introduced and developed a method for qualitative analysis of oriented texture patterns; the method depends on the association of an image representing an oriented texture pattern with the appearance of a phase-portrait diagram and the corresponding parameters of the system of differential equations.

The method of linear phase-portrait analysis is described with illustrative examples in the following sections. The phase-portrait method was used in the present study for the detection of potential sites of architectural distortion in prior mammograms.

3.3.1   PHASE PORTRAITS

The phase-portrait diagram is an analytical method for studying systems of first-order differential equations [114]. Let $p(t)$ and $q(t)$, $t \in \mathbb{R}$, be two differentiable functions of time t, related by a system of first-order differential equations as

$$
\begin{aligned}
\dot{p}(t) &= F[p(t), q(t)], \\
\dot{q}(t) &= G[p(t), q(t)],
\end{aligned}
\tag{3.3}
$$

where $\dot{p}(t)$ and $\dot{q}(t)$ represent the first-order derivatives with respect to time, and F and G are two functions of p and q [64]. The solution $(p(t), q(t))$ to Equation 3.3, with given initial conditions $p(0)$ and $q(0)$, can be considered as a parametric trajectory of a hypothetical particle in the pq plane, placed at $(p(0), q(0))$ at time $t = 0$, and moving through the pq plane with velocity $(\dot{p}(t), \dot{q}(t))$.

The pq plane is commonly referred to as the *phase plane* of the system of first-order differential equations [88]. The path followed by the hypothetical particle is called a *streamline* of the vector (velocity) field $(\dot{p}, \dot{q})$. The *phase portrait* is a graph of the possible streamlines in the phase plane, and a *fixed point* or *singular point* of Equation 3.3 is a point in the phase plane where $\dot{p}(t) = 0$ and $\dot{q}(t) = 0$; a particle left at a fixed point remains stationary for all values of t.

If the system of first-order differential equations is linear and affine, Equation 3.3 can be represented by

$$\begin{pmatrix} \dot{p}(t) \\ \dot{q}(t) \end{pmatrix} = \mathbf{A} \begin{pmatrix} p(t) \\ q(t) \end{pmatrix} + \mathbf{b}, \qquad (3.4)$$

where $\mathbf{A}$ is a 2×2 matrix and $\mathbf{b}$ is a 2×1 column matrix (a vector). For a linear and affine system, there are only three types of phase portraits: node, saddle, and spiral [114]. The type of phase portrait can be determined from the nature of the eigenvalues of $\mathbf{A}$ [23, 60, 64, 88, 89], as shown in Table 3.1.

The eigenvalues of $\mathbf{A}$ are given by

$$\lambda_1 = \frac{tr(\mathbf{A})}{2} + \frac{\sqrt{[tr(\mathbf{A})]^2 - 4\,det(\mathbf{A})}}{2} \qquad (3.5)$$

and

$$\lambda_2 = \frac{tr(\mathbf{A})}{2} - \frac{\sqrt{[tr(\mathbf{A})]^2 - 4\,det(\mathbf{A})}}{2}, \qquad (3.6)$$

where $tr(\mathbf{A})$ is the trace of matrix $\mathbf{A}$ and $det(\mathbf{A})$ is the determinant of $\mathbf{A}$.

The center (p_0, q_0) of the phase portrait is given by the fixed point of Equation 3.4:

$$\begin{pmatrix} \dot{p}(t) \\ \dot{q}(t) \end{pmatrix} = 0 \Rightarrow \begin{pmatrix} p_0 \\ q_0 \end{pmatrix} = -\mathbf{A}^{-1}\mathbf{b}. \qquad (3.7)$$

Solving Equation 3.4 yields a linear combination of complex exponentials for $p(t)$ and $q(t)$; the exponents are represented by the eigenvalues of $\mathbf{A}$ multiplied by the time variable t. Table 3.1 illustrates the streamlines obtained by solving Equation 3.4 for a node, a saddle, and a spiral phase portrait: the solid lines indicate the movement of the $p(t)$ and the $q(t)$ components of the solution, and the dashed lines indicate the streamlines [23]. The formation of each phase portrait type is described as follows:

- *Node*—If the components $p(t)$ and $q(t)$ are exponentials and they either simultaneously converge to, or diverge from, the fixed-point coordinates p_0 and q_0, the generated pattern is known as a node. A node pattern is obtained when the eigenvalues of $\mathbf{A}$ have the same sign; from Equations 3.5 and 3.6, a node pattern is generated if

$$0 < det(\mathbf{A}) < \frac{[tr(\mathbf{A})]^2}{4}.$$

- *Saddle*—If the components $p(t)$ and $q(t)$ are exponentials and one of the components (either $p(t)$ or $q(t)$) converges to the fixed point and the other diverges, a saddle pattern is obtained. A saddle pattern occurs when the eigenvalues of $\mathbf{A}$ are real and have opposite signs; this condition is achieved if

$$det(\mathbf{A}) < 0.$$

Table 3.1: Phase portraits for a system of linear first-order differential equations. Solid lines indicate the movement of the $p(t)$ and the $q(t)$ components of the solution; dashed lines indicate the streamlines. Reproduced with permission from: F. J. Ayres and R. M. Rangayyan. "Characterization of architectural distortion in mammograms." IEEE Engineering in Medicine and Biology Magazine, 24(1):59–67, January 2005. © IEEE.

Phase portrait type	Eigenvalues	Streamlines	Appearance of the orientation field
Node	Real eigenvalues of same sign		
Saddle	Real eigenvalues of opposite sign		
Spiral	Complex eigenvalues		

- *Spiral*—If the components $p(t)$ and $q(t)$ are exponentially modulated sinusoidal functions, the resulting streamline forms a spiral curve. In this case, the eigenvalues of **A** form a pair of complex conjugate numbers with nonzero imaginary parts; this condition occurs if

$$det(\mathbf{A}) > \frac{[tr(\mathbf{A})]^2}{4}.$$

3.4 ANALYSIS OF ORIENTATION FIELDS USING PHASE PORTRAITS

The model in Equation 3.4 can be used to perform qualitative analysis of orientation fields, as proposed by Rao and Jain [89]. Consider the vector field model:

$$\vec{v} = \begin{pmatrix} v_x \\ v_y \end{pmatrix} = \mathbf{A} \begin{pmatrix} x \\ y \end{pmatrix} + \mathbf{b} , \tag{3.8}$$

where

$$\mathbf{A} = \begin{bmatrix} a & b \\ c & d \end{bmatrix}, \quad \mathbf{b} = \begin{bmatrix} e \\ f \end{bmatrix} . \tag{3.9}$$

The vector $\vec{v}$ is an affine function of the coordinates (x, y). A particle on the image plane moving with the velocity $\vec{v}(x, y)$ will follow a trajectory analogous to the time evolution of the dynamical system in Equation 3.4. Comparing Equation 3.8 to Equation 3.4, the vector $\vec{v}$ can be associated with the state velocity $(\dot{p}(t), \dot{q}(t))$, and the position (x, y) of the particle can be associated with the state $(p(t), q(t))$. Then, the *orientation field* produced by Equation 3.8 can be defined as

$$\phi(x, y | \mathbf{A}, \mathbf{b}) = \tan^{-1} \left(\frac{v_y}{v_x} \right) , \tag{3.10}$$

where $\phi(x, y | \mathbf{A}, \mathbf{b})$ is the angle of the vector $\vec{v}$ with the x axis.

Table 3.1 illustrates the three types of phase portraits and the corresponding orientation fields generated by a system of linear first-order differential equations. The orientation field of an image $I(x, y)$ with oriented texture can be obtained using the real Gabor filter bank described in Section 3.2.1. Let $M(x, y)$ and $\theta(x, y)$ be the magnitude and angle components of the orientation field of $I(x, y)$. At every pixel location (x, y), let the error between $\theta(x, y)$ and a synthetic orientation field $\phi(x, y | \mathbf{A}, \mathbf{b})$ be defined as

$$r(x, y) = \sin (\theta(x, y) - \phi(\vec{x}, y | \mathbf{A}, \mathbf{b})) . \tag{3.11}$$

The *sine* of the difference between angles, as shown in Equation 3.11, serves as a better measure of the difference between two orientations than the difference between the angles [88]. It should be noted that a phase portrait is a qualitative representation of an orientation field, but not a vector field; the use of the *sine* of the difference between angles removes the ambiguity that arises if the sign of the orientation is considered (e.g., lines oriented at $0°$ and $180°$ have opposite signs). For example, if two line segments are oriented at $1°$ and $179°$ with respect to the x-axis, the difference between the angles is $178°$, whereas the *sine* of the difference is close to zero; the latter measure accurately represents the fact that the line segments possess nearly the same orientation. Then, the sum of the squared errors or the error measure, weighted by the magnitude of the orientation field, is given by

$$\epsilon^2(\mathbf{A}, \mathbf{b}) = \sum_x \sum_y M(x, y) \sin^2(\theta(x, y) - \phi(x, y | \mathbf{A}, \mathbf{b})) . \qquad (3.12)$$

Minimizing $\epsilon^2(\mathbf{A}, \mathbf{b})$ with respect to the elements of $\mathbf{A}$ and $\mathbf{b}$ yields a set of optimal parameters $\mathbf{A}_{opt}$ and $\mathbf{b}_{opt}$; the parameters are related to the synthetic orientation field that makes the best approximation of the orientation field of $I(x, y)$. The parameters $\mathbf{A}_{opt}$ and $\mathbf{b}_{opt}$ facilitate the determination of the type of phase portrait and the location of the fixed point of $\phi(x, y | \mathbf{A}_{opt}, \mathbf{b}_{opt})$ for providing a qualitative description of the oriented texture pattern and its focal point, respectively, in the image under investigation. For the minimization of $\epsilon^2(\mathbf{A}, \mathbf{b})$, the simulated annealing method [115] has been used.

Ayres [23] and Ayres and Rangayyan [116, 117] conducted studies on several optimization procedures and concluded that the application of simulated annealing followed by the nonlinear least-squares method [118] could be the appropriate optimization strategy when a high level of noise is present. On the other hand, if the oriented texture patterns are clearly evident and highly coherent (i.e., the orientation field has a local dominant orientation at every pixel) in a given image, a faster optimization strategy, such as the combination of the iterative linear least-squares and the nonlinear least-squares methods, could be used [116, 117]. However, simulated annealing [115] followed by a nonlinear least-squares algorithm [118] was used in the present work for robustness. The nonlinear least-squares algorithm used in this work is the Levenberg-Marquardt method [119]; the procedure was implemented using a collection of routines for numerical computation in the GNU Scientific Library [120].

The simulated annealing algorithm [115] is a global optimization procedure for the localization of a good approximation to the global minimum of a given function in a large search space. At each iteration, from the current parameter vector, a new estimate of the optimal parameters is computed by taking a random step of limited length in the parameter space. The new parameter vector is accepted as a new estimate of the optimal parameters if it leads to an improved solution (i.e., the cost function to be minimized is smaller for the new parameter vector than that for the current parameter vector). If the new parameters represent a worse solution (i.e., the cost function to be minimized is larger for the new parameter vector than that for the current parameter vector), the parameters may be accepted with a probability that decreases with the number of iterations; these probabilities ultimately lead the system to move to states of lower energy. Allowing the algorithm to move to a worse solution prevents the algorithm from getting stuck in a local minimum. If the probability of acceptance of a worse solution is decreased as the iterations progress, the algorithm will converge to the global optimum of the function being optimized [23]. Typically, the steps are repeated until the system reaches a state that is adequate for the application of interest, or until a given computation budget has been exhausted.

Large orientation fields of heterogeneous images could contain complex patterns that are generated by superpositions of several patterns or overlapping structures; this may result in the presence of multiple focal points [23, 88]. The approximation of real orientation fields using Equation 3.10 is valid for local analysis only; a reasonable strategy to extend the method for local analysis of orientation fields to the analysis of a large orientation field is through the analysis of the large orientation field at multiple locations (within a small window), and to accumulate the acquired information in a form that would help in the identification of the various significant patterns or structures present in the overall orientation field.

Rao and Jain [89] proposed the following method for the analysis of large orientation fields:

1. Create three images of the same size as that of the image or orientation field under analysis; the images are referred to as *phase-portrait maps*. Initialize the three phase-portrait maps to zero.

2. Perform analysis by moving a small sliding window throughout the orientation field. For every position of the analysis window, perform the following steps:

   (a) Use the local analysis procedure described above to find the optimal parameters $\mathbf{A}_{opt}$ and $\mathbf{b}_{opt}$ that give the best representation of the orientation field within the analysis window.

   (b) Determine the type of phase portrait and the fixed-point location associated with the orientation field within the analysis window from $\mathbf{A}_{opt}$ and $\mathbf{b}_{opt}$.

   (c) Select the phase-portrait map corresponding to the phase-portrait type determined above and increment the value present at the pixel nearest to the fixed-point location. This procedure can be referred to as *vote casting*.

When all votes are cast, the phase-portrait maps could be analyzed to detect the presence of texture patterns in the given image or orientation field [23]. If a part of the orientation field consists of orientations converging to or diverging from a central point, in a manner similar to a node pattern, it is expected that the node map will contain a large number of accumulated votes close to the geometrical focal point of the observed pattern. In a similar manner, the presence of patterns that resemble saddle or spiral patterns will lead to the accumulation of a number of votes in the corresponding phase-portrait maps. This procedure has been shown to facilitate the detection of node patterns and architectural distortion in mammograms [38, 60, 63, 64].

3.5   ILLUSTRATIVE EXAMPLE

In order to illustrate the technique described in Section 3.4, a test image exhibiting oriented patterns was used and is displayed in part (a) of Figure 3.2: the image includes three prominent node patterns.

The magnitude $M(x, y)$ and angle $\theta(x, y)$ components of the orientation field of $I(x, y)$ were obtained using the real Gabor filter bank described in Section 3.2.1. The filter bank is composed of 180 filters, spanning the range of orientations $0°$ to $179°$ in steps of $1°$. The orientation field was analyzed using the procedure for large orientation fields presented in Section 3.4.

Figure 3.2: Analysis of an image of fireworks with prominent node patterns: (a) original image of size 528×362 pixels; (b) magnitude response image; (c) orientation field superimposed on the original image, with needles drawn for every 12th pixel; (d) node map, range [0, 275]; (e) saddle map, empty; (f) spiral map, range [0, 7]. Gabor parameters: $\tau = 1$ pixel, $l = 8$; the phase-portrait analysis window was of size 50×50 pixels.

The Gabor magnitude and angle response as well as the phase-portrait maps are presented in Figure 3.2. For the image shown in part (a) of Figure 3.2, the parameters of the Gabor filters were selected to be: $\tau = 1$ pixel, $l = 8$; the size of the analysis window for the phase-portrait procedure was 50×50 pixels, which was slid one pixel per step. The orientation field was downsampled by a factor of four, using the procedure described in Section 4.1.4, before the application of phase-portrait analysis.

It can be observed that there is a concentration of votes in the node map at the focal points of the three firework patterns. It can also be noticed that votes have been cast in the spiral map although the test image contains no clearly evident spiral patterns; various components of multiple patterns in a given image could combine to present partial representations of other types of patterns than the dominant or most obvious pattern. Nevertheless, the miscast votes have been spread across the spiral map and do not result in the presence of large peaks in the maps that are as prominent as the peaks associated with the correct phase-portrait map corresponding to the dominant pattern, that is, node. No vote was cast for the saddle patterns and the saddle map is empty.

3.6   REMARKS

In this chapter, methods for the detection and analysis of oriented patterns using Gabor filters and phase-portrait analysis were discussed. The results of filtering using a bank of 180 Gabor filters are combined to form two maps: the magnitude map, which indicates the intensity of the oriented feature at each pixel, and the angle map, which indicates the orientation of the feature at each pixel. The magnitude and angle maps are together referred to as the orientation field of the image under analysis. The real Gabor filter is recommended when high detection performance and angular accuracy are required, as in the case of noisy images or when the presence of oriented features is subtle [23]. The general methodology for the analysis of oriented texture using phase portraits, originally developed by Rao and Jain [89], as well as the optimization procedure and the application of the method to a set of test images, were also presented in this chapter.

Gabor filters and the phase-portrait analysis are used in the present work for the detection of potential sites of architectural distortion in prior mammograms of interval-cancer cases, as shown in Chapter 4.

CHAPTER 4

Detection of Potential Sites of Architectural Distortion

4.1 DETECTION OF ARCHITECTURAL DISTORTION

In the present work, potential sites of architectural distortion in prior mammograms of interval-cancer cases were detected initially by the analysis of oriented texture patterns with the application of Gabor filters and linear phase-portrait models [38, 39, 84, 87]. The method for the detection of potential sites of architectural distortion involves the following stages: approximate segmentation of the breast portion in a given mammogram, use of a bank of 180 real Gabor filters with angles spaced evenly over the range $[-\pi/2, \pi/2]$ to obtain the magnitude response and orientation field, selection of the CLS of interest (i.e., spicules and fibroglandular tissue), filtering and downsampling of the core CLS pixels, and application of the linear phase-portrait modeling procedure with specific conditions applied to the filtered orientation field to yield two types of phase-portrait maps: node and saddle [60, 64]. The node map, obtained through the application of phase-portrait analysis, was further analyzed to detect peaks related to potential sites of architectural distortion; however, the procedure also resulted in the detection of a number of FP sites [39]. The flowchart shown in Figure 4.1 summarizes the procedure described above. The methods presented in this chapter were developed by Ayres et al. [23], and were used in the present work with some modifications. The methods are discussed in detail in the following subsections.

4.1.1 SEGMENTATION OF THE BREAST PORTION IN A MAMMOGRAM

At first, each mammographic was filtered using a Gaussian filter (with a standard deviation of 2 pixels and size of 13×13 pixels at the resolution of 50 μm/pixel and 12 bits/pixel) and downsampled to 200 μm/pixel and 8 bits/pixel resolution. The approximate breast portion of a given mammogram was segmented by applying Otsu's thresholding method [121]. The morphological opening filter [122] with a disk-shaped structuring element of radius 25 pixels (5 mm at 200 μm/pixel) was used for smoothing the edges [64].

Although improved methods for the delineation of the breast boundary using active contour models have been proposed by Ferrari et al. [123], a simpler approach as above is adequate for the present application [23]. Pixels outside the breast area were removed from further consideration and the method for the detection of architectural distortion was applied to the segmented breast portion in the given mammogram.

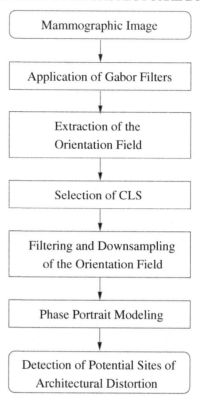

Figure 4.1: Summary of the procedure for initial detection of potential sites of architectural distortion in prior mammograms. Reproduced with permission from: S. Banik, R. M. Rangayyan, and J. E. L. Desautels. "Digital Image Processing and Machine Learning Techniques for the Detection of Architectural Distortion in Prior Mammograms." In K. Suzuki, Ed., *Machine Learning in Computer-aided Diagnosis: Medical Imaging Intelligence and Analysis*, pp. 24–63, IGI Global, Hershey, PA, January 2012. © IGI Global.

4.1.2   EXTRACTION OF THE ORIENTATION FIELD

For each mammographic image, a magnitude response and an orientation field were obtained by using the magnitude response and angle of the Gabor filter with the highest response at each pixel. Because a Gabor filter has a nonzero magnitude response at the origin of the frequency plane (DC or zero frequency) [23], the low-frequency components of the mammographic image could influence the result of the Gabor filter. Although such an influence will not have any impact on the computation of the orientation field angle (the same influence will appear at all angles), the magnitude response will be modified and exhibit values that are affected by the low-frequency content of the image. It is desirable to reduce the effect of the low-frequency components of the mammographic image in the orientation field magnitude as they are not related to the presence of architectural distortion.

For this reason, prior to the extraction of the orientation field, the mammographic images were high-pass filtered by computing the difference between the original mammographic image and a low-pass-filtered version of the same image. The low-pass filter used in this step is a Gaussian filter with $\sigma_{LPF} = \sigma_y$ (as defined in Section 3.2.1) with unit gain at the origin of the frequency plane.

4.1.3   SELECTION OF CURVILINEAR STRUCTURES

The identification of CLS is an important factor in the detection of abnormalities in mammograms [23, 58, 124]. Mammograms could contain many CLS corresponding to blood vessels, milk ducts, parenchymal tissue, ligaments, and edges of the pectoral muscle. The presence of certain types of CLS could be used in the characterization of some particular lesions in mammograms, such as spiculated masses [58, 125, 126] and architectural distortion [60, 62, 64], or the asymmetric arrangement of the oriented texture in the breast image [110]. On the other hand, some lesions (e.g., circumscribed masses) could be obscured by several CLS superimposed on the lesions; the altered appearance of the lesions could lead to FN detection and misdiagnosis. Therefore, efficient and accurate detection and classification of CLS could help in enhancing the performance of CAD algorithms [23].

Evans et al. [127] presented a method for statistical characterization of normal CLS in mammograms: six shape features were computed from automatically detected CLS. Using principal component analysis (PCA), the first two dimensions were modeled using a Gaussian mixture model. Wai et al. [128] proposed a method for the segmentation of CLS based on physical modeling of CLS in the breast; qualitative experiments suggested that the method produced robust and well-localized responses in the presence of noise. Zwiggelaar et al. [124] studied the performance of different methods for the detection and classification of CLS in mammograms; the methods included the use of a line operator [129], oriented bins [130], steerable filters [106], and ridge detectors [131]. The use of line operators for CLS detection yielded the best result with $A_z = 0.94$. Cross-sectional analysis of the detected profiles and the use of PCA for dimensionality reduction resulted in good discrimination between spicules and ducts ($A_z = 0.75$).

Although the Gabor filter bank used in the present work is sensitive to linear structures, such as spicules and fibers, it also identifies other strong edges present in the given mammographic image, such as edges of the pectoral muscle, edges of the parenchymal tissue, and vessel walls, as oriented features. Strong edges around the fibroglandular disk [132] could be helpful in the detection of a particular form of architectural distortion known as focal retraction [66]. However, only linear structures related to fibroglandular tissues are important and need to be identified as oriented features in order to detect architectural distortion. Detection and analysis of the CLS present in mammograms could improve the performance of algorithms for the detection of stellate patterns, such as spiculated masses and architectural distortion [124]. The method for the selection of CLS in the present work is composed of three stages: segmentation of the breast area, detection of core CLS pixels, and rejection of CLS pixels at sites with a strong gradient [62].

The method for segmentation of the breast area in the given mammographic image is described in Section 4.1.1. The detection of core CLS pixels was performed by applying the nonmaximal suppression (NMS) technique [133] to the Gabor magnitude image. In the NMS algorithm, the core CLS pixels are identified by comparing each pixel in the magnitude image with its neighbors along the direction perpendicular to the local orientation field angle; see Figure 4.2 [23]. A pixel in considered to be a core CLS pixel if it possesses larger magnitude value than the corresponding neighbors. NMS is widely used in many edge detectors (e.g., the Canny edge detector [134]).

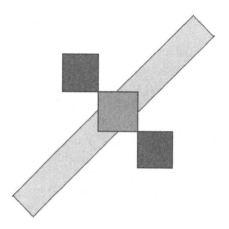

Figure 4.2: The NMS technique: the elongated rectangle (in gray) denotes the presence of a CLS, whereas the squares denote pixels along a direction perpendicular to the CLS orientation. The central green square indicates a core CLS pixel.

In the final step, the presence of a strong gradient was assumed to cause a ripple in the magnitude image resulting in an erroneous detection of a CLS. The core CLS pixels associated with the strong gradients were rejected using the rejection criteria proposed by Karssemeijer and te Brake [58] in the context of the detection of stellate lesions. The gradient of the mammographic image was computed by filtering with the first derivative of a Gaussian with a standard deviation of five pixels (1 mm). The direction of the gradient was compared to the direction of the orientation field for each core CLS pixel; if the difference between the direction of the orientation field and the direction perpendicular to the gradient was less than $\pi/6$, the core CLS pixel under consideration was rejected.

The CLS within the fibroglandular disk typically exhibit reduced contrast as compared to similar CLS outside the fibroglandular disk [23]. Consequently, the CLS within the fibroglandular disk will have smaller magnitude values than those of CLS outside the fibroglandular disk. In order to assign the same weight to all CLS pixels independent of their location, and to ensure that relevant CLS with low contrast, such as spicules within the fibroglandular disk, are not missed, the magnitude image $M(x, y)$ was replaced by an image composed of only core CLS pixels, $M_{CLS}(x, y)$, defined as follows:

$$M_{CLS}(x, y) = \begin{cases} 1 & \text{if the pixel at } (x, y) \text{ is a core CLS pixel} \\ 0 & \text{otherwise.} \end{cases} \qquad (4.1)$$

The image $M_{CLS}(x, y)$ provides important information regarding the presence and arrangement of CLS. Because the presence of architectural distortion is represented by the geometrical arrangement of the associated CLS rather than their intensity, the magnitude of the detected CLS is of lesser importance. Figures 4.3 and 4.4 show the results of CLS selection with a full mammogram and an ROI, respectively.

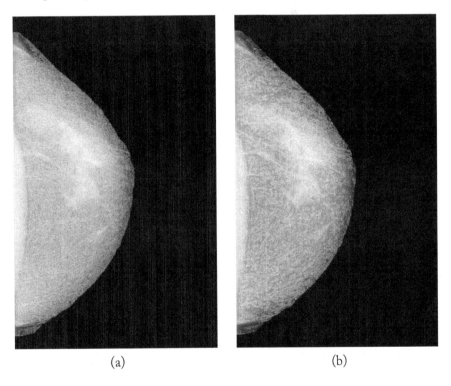

(a)   (b)

Figure 4.3: NMS and CLS results overlaid on a full mammographic image. (a) NMS results. (b) CLS results (after CLS rejection); the pixels marked in white correspond to CLS pixels that are retained for further analysis.

4.1.4   FILTERING AND DOWNSAMPLING THE ORIENTATION FIELD

The orientation field was filtered and downsampled in order to reduce noise and the computational cost associated with the processing of full mammograms. Let $h(x, y)$ be a Gaussian filter of standard

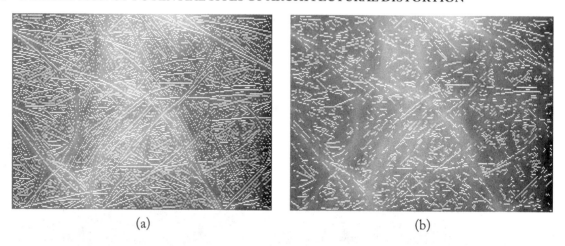

(a)   (b)

Figure 4.4: NMS and CLS results overlaid on a mammographic ROI. (a) NMS results. (b) CLS results (after CLS rejection); the pixels marked in white correspond to CLS pixels that are retained for further analysis.

deviation σ_f, defined as

$$h(x, y) = \frac{1}{2\pi\sigma_f^2} \exp\left[-\frac{1}{2}\left(\frac{x^2 + y^2}{\sigma_f^2}\right)\right]. \tag{4.2}$$

Let $s(x, y) = M_{CLS}(x, y) \sin[2\theta(x, y)]$ and $c(x, y) = M_{CLS}(x, y) \cos[2\theta(x, y)]$, where $\theta(x, y)$ is the orientation field angle and $M_{CLS}(x, y)$ is the magnitude field composed of only core CLS pixels as defined in Equation 4.1. In the present work, $\sigma_f = 7$ pixels was used. The orientation field magnitude after CLS selection was obtained as

$$M(x, y) = (h * M_{CLS})(x, y), \tag{4.3}$$

where the asterisk denotes convolution. Then, the filtered orientation field angle $\theta_f(x, y)$ was computed as

$$\theta_f(x, y) = \frac{1}{2} \tan^{-1}\left(\frac{(h * s)(x, y)}{(h * c)(x, y)}\right). \tag{4.4}$$

The filtered orientation field was downsampled by a factor of four, resulting in the downsampled orientation field θ_d as

$$\theta_d(x, y) = \theta_f(4x, 4y). \tag{4.5}$$

The filtered and downsampled angle and magnitude images have a resolution of 0.8 mm/pixel.

4.1.5 ESTIMATING THE PHASE-PORTRAIT MAPS

To facilitate the detection of spiculated distortion [23, 60, 64], a penalty term was included in the estimation process of the phase-portrait maps; the penalty term is represented by

$$\epsilon^2_{\text{node}}(\mathbf{A}) = 1000 \left[(a-d)^2 + 4bc \right]^2 , \tag{4.6}$$

where

$$\mathbf{A} = \begin{bmatrix} a & b \\ c & d \end{bmatrix} , \tag{4.7}$$

which accounts for a higher penalty (cost) for any deviation in configurations of the matrix $\mathbf{A}$ from a spiculated node appearance. Therefore, the sum of the squared error is given by

$$\epsilon^2(\mathbf{A}, \mathbf{b}) = \sum_x \sum_y M(x, y) \sin^2[\theta(x, y) - \phi(x, y|\mathbf{A}, \mathbf{b})] + \epsilon^2_{\text{node}}(\mathbf{A}) , \tag{4.8}$$

where the range of the summation indices x and y is the set of the pixel locations within the analysis window. Estimates of $\mathbf{A}$ and $\mathbf{b}$ that minimize $\epsilon^2(\mathbf{A}, \mathbf{b})$ were obtained as follows [23]:

1. Initial estimates $\mathbf{A}_{SA}$ and $\mathbf{b}_{SA}$ of $\mathbf{A}$ and $\mathbf{b}$ were obtained through the minimization of $\epsilon^2(\mathbf{A}, \mathbf{b})$ using the simulated annealing [115] algorithm described in Section 3.4.

2. The optimal estimates $\mathbf{A}_{opt}$ and $\mathbf{b}_{opt}$ were obtained by refining the estimates $\mathbf{A}_{SA}$ and $\mathbf{b}_{SA}$ using a nonlinear least-squares algorithm [118].

4.1.6 SHAPE-CONSTRAINED PHASE-PORTRAIT MODEL

Using the optimal values $\mathbf{A}_{opt}$ and $\mathbf{b}_{opt}$, the type of phase portrait was determined by the eigenvalues of $\mathbf{A}_{opt}$ and the fixed-point location was obtained by Equation 3.7. A vote was cast at the location of the fixed point in the corresponding phase-portrait map; the procedure [60, 135] is described in the following paragraphs.

Adoption of a symmetric matrix A
The general model in Equation 3.8 could lead to synthetic orientation fields that are inappropriate for the specific practical application; if the inner angle between the eigenvectors of $\mathbf{A}$ is small, or if the eigenvalues of $\mathbf{A}$ are significantly different in magnitude, the orientation field generated by the phase-portrait model may generate a pattern composed of almost parallel lines [23]. Consequently, the model may produce a phase-portrait pattern without a fixed point (any specific geometric focus).

In order to prevent the eigenvectors from having small inner angles, the matrix $\mathbf{A}$ in Equation 3.9 was replaced by a symmetric matrix model to impose orthogonality of the eigenvectors [60], as follows:

$$\mathbf{A} = \begin{bmatrix} a & b \\ b & c \end{bmatrix}. \tag{4.9}$$

The symmetric matrix model ensures that the eigenvectors of $\mathbf{A}$ are mutually orthogonal and the eigenvalues of $\mathbf{A}$ are real-valued [136], and results in only node and saddle phase portraits. Spiral patterns are not expected to appear in mammograms with architectural distortion and are not of interest in the present work.

Vote casting and detection

In order to prevent the matrix $\mathbf{A}$ from having significantly different eigenvalues, a restriction can be imposed on the condition number (i.e., the ratio of the largest to the smallest singular values) of the matrix. The condition number is related to the degree of numerical precision achievable in the process of inverting the matrix and is always greater than or equal to one. In the case of a symmetric matrix, the singular values are equal to the absolute values of the eigenvalues. A high condition number of a given matrix suggests that the corresponding inverse matrix will be highly sensitive to small perturbations in the original matrix and the geometrical focus of the pattern (represented by the fixed-point location) cannot be determined accurately. A singular matrix refers to an infinite condition number and indicates a phase-portrait pattern without a fixed point. Therefore, in practical applications, a high condition number indicates the absence of a significant geometrical focus for a given orientation field.

In the vote-casting step, inappropriate phase-portrait models are rejected based on the condition number of $\mathbf{A}$ [60]; if the condition number is greater than 3.0, the corresponding phase-portrait model is considered to be unsuitable for further analysis. The limit on the condition number prevents numerical instabilities in the computation of $\mathbf{A}^{-1}$, and ensures that the precision in the determination of the fixed-point location will not be affected by ill-conditioning of $\mathbf{A}$. In addition, the constraint facilitates robust detection of an accurate fixed-point location.

At each position of the moving analysis window, a fixed-point location was obtained for the estimated phase-portrait model; the results for the analysis window under consideration were accepted or rejected based on the distance from the center of the analysis window to the fixed-point location. If the associated distance was less than three pixels (2.4 mm) or greater than 20 pixels (16 mm), the results for the current analysis window were rejected.

Regions of architectural distortion are not expected to contain a high-density area or mass at the center of the pattern [13]; in addition, the typical geometrical patterns associated with architectural distortion may not be present precisely at the center of the region. Regardless, the central regions of sites of architectural distortion could contain patterns related to other structures in the mammogram; the lower limit of 2.4 mm on the distance to the fixed point would prevent FPs due to such interfering patterns. On the other hand, if the distance from the center of the analysis window to the fixed-point location is large with respect to the size of the analysis window, the fixed-point location cannot be accurately determined; the upper limit of 16 mm is used to address this issue.

If both conditions for the suitability of the phase portrait, i.e., the condition number and the distance to the fixed point, were satisfied, a vote was cast at the fixed-point location in one of the two phase-portrait maps: node or saddle. The magnitude of the vote was set equal to the ratio of the measure of fit $\epsilon^2(\mathbf{A}, \mathbf{b})$ (defined in Equation 3.12) to the condition number of $\mathbf{A}$ to emphasize the isotropy of the phase portrait [23]. The process described above was repeated for every position of the analysis window.

4.2   POTENTIAL SITES OF ARCHITECTURAL DISTORTION

The votes related to the presence of a pattern of architectural distortion may be scattered over a small region, rather than being grouped exactly at the same location in the corresponding phase-portrait map. For this reason, the node map was filtered with a Gaussian window of size 35×35 pixels with the empirically determined value of $\sigma = 6$ pixels (4.8 mm) in order to group votes placed in close proximity to one another. The node map was analyzed to detect peaks that were expected to be related to sites of architectural distortion. In most cases, the saddle maps were observed to contain scattered responses, and lacked discrimination across the mammograms analyzed; hence they were removed from further consideration.

The results of application of the methods are illustrated in Figures 4.5 through 4.8 for two prior mammograms of two interval-cancer cases. The rectangles indicate the areas of architectural distortion identified by the radiologist (J.E.L.D.); see Section 5.1 for details on the datasets used. The two mammograms, the Gabor magnitude responses, the orientation fields, and the node maps along with the corresponding zoomed areas of architectural distortion are illustrated in Figures 4.5 through 4.8. In parts (a) and (b) of Figure 4.8, the most dominant peaks are evident within the sites of architectural distortion, as expected.

4.3   REMARKS

In this chapter, the methods used for the detection of potential sites of architectural distortion based on Gabor filters and linear phase-portrait analysis were presented. The methods of NMS, CLS selection, constrained phase-portrait modeling with a symmetric matrix $\mathbf{A}$, and incorporation of the condition number of $\mathbf{A}$ into the vote-casting procedure were described in detail. The methods presented in this chapter were developed by Ayres et al. [23], and were used in the present work with some modifications as an important initial step for the detection of potential sites of architectural distortion in prior mammograms of interval-cancer cases. The datasets, methodology used, and results obtained in the present work are presented in the following chapters.

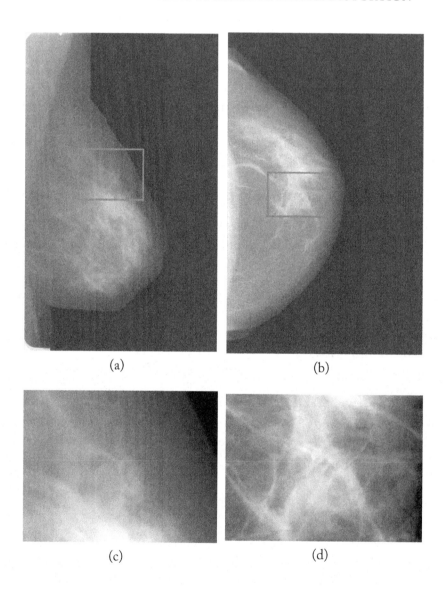

Figure 4.5: Examples of prior mammograms of interval-cancer cases. (a) Image size: 1370×850 pixels at 200 μm/pixel. The rectangle is of size 57.5 mm $\times$ 43.5 and indicates the region of architectural distortion identified by a radiologist. (b) Image size: 1376×850 pixels at 200 μm per pixel. The rectangle is of size 51.3 mm $\times$ 37.6 mm and indicates the region of architectural distortion identified by a radiologist. (c) Zoomed view of the rectangular area shown in part (a). (d) Zoomed view of the rectangular area shown in part (b).

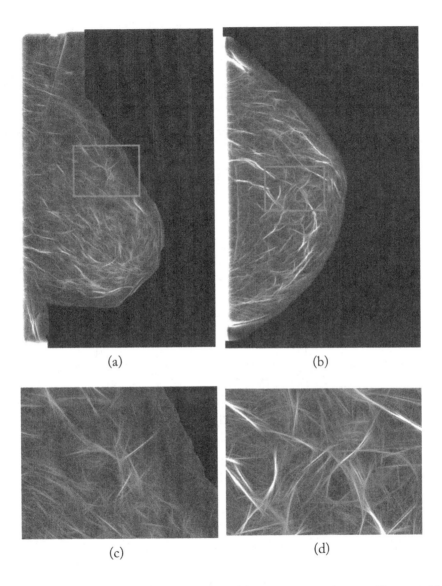

(a)   (b)

(c)   (d)

Figure 4.6: Examples of Gabor magnitude responses: (a) for the image shown in Figure 4.5 (a); (b) for the image shown in Figure 4.5 (b); (c) zoomed view of the rectangular area shown in part (a); (d) zoomed view of the rectangular area shown in part (b).

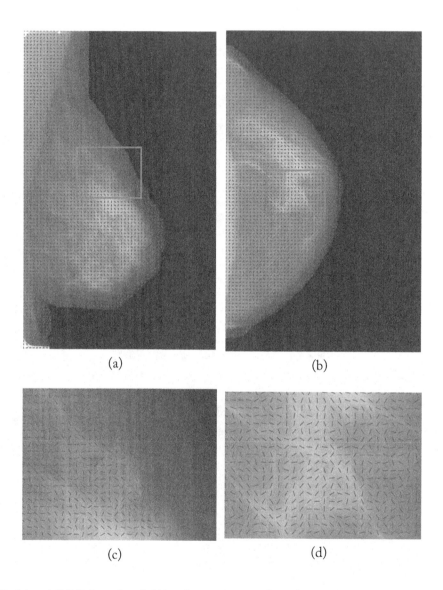

(a)

(b)

(c)

(d)

Figure 4.7: (a) and (b) Orientation field angle superimposed on the mammographic images shown in Figure 4.5 (a) and (b), respectively. Needles are drawn for every 15th pixel. (c) and (d) Zoomed views of the rectangular areas shown in part (a) and (b), respectively; needles are drawn for every 10th pixel for clarity.

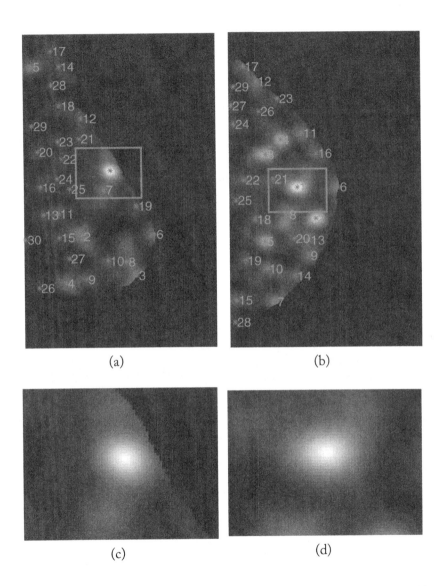

Figure 4.8: (a) and (b) The node maps at 800 μm/pixel for the mammograms shown in Figure 4.5 (a) and (b), respectively. Each asterisk mark (*) corresponds to a peak position detected automatically in the node map. The numbers next to the asterisk marks indicate the peaks in descending order of magnitude. (c) and (d) Zoomed views of the rectangular areas shown in part (a) and (b), respectively.

CHAPTER 5

Experimental Set Up and Datasets

5.1 DATASETS OF MAMMOGRAMS

Mammographic images were obtained from a database of 1,745 digitized mammograms of 170 subjects from Screen Test: Alberta Program for the Early Detection of Breast Cancer [137, 138]. Ethics approval for the study was obtained from the Conjoint Health Research Ethics Board, Office of Medical Bioethics, University of Calgary, and the Calgary Regional Health Authority. The film mammograms were digitized at the spatial resolution of 50 μm and gray-scale resolution of 12 bits per pixel (bpp) using the Lumiscan 85 laser scanner (Lumisys, Sunnyvale, CA), resulting in uncompressed raw images of between 40 and 60 MB each [137].

The radiologist (Dr. J.E.L. Desautels) who analyzed and annotated the images used in the present study has more than 40 years of experience in mammography, of which more than 20 years is in screening for breast cancer; he was also a member of the team of radiologists in the Screen Test Program and interpreted the mammograms at the original instances of screening. All cases of interval cancer were reviewed by a panel of five experienced radiologists in the screening program as part of the standard protocol. The laterality, location, and nature of architectural distortion and/or other signs of breast cancer were determined by the radiologists and pathologists involved in the diagnostic imaging and other investigations.

Two datasets of prior mammographic images were used in the present study. The first dataset includes prior mammographic images of interval-cancer cases, with no related detection or diagnostic mammograms available, and normal control cases. In order to analyze the effect of cross-validation with other detasets, a second dataset of screen-detected cancer cases including detection and prior mammograms was used. The details of the datasets are described in the following sections.

5.1.1 INTERVAL CANCER

In consultation with the radiologist, all of the prior mammograms of interval-cancer cases available in the database have been included in the study, except six images in which no suspicious parts could be identified. The corresponding detection mammograms were not available for the present study. The dataset for this study includes a total of 106 prior mammographic images of 56 individuals diagnosed with cancer in a single breast in each case. All but two of the 106 prior mammograms had been declared to be free of any sign of breast cancer at the time of their original acquisition and interpretation in the screening program; the other two cases had been referred for biopsy although no definite signs of malignancy were present. The time interval between the detection mammograms (not available) and prior mammograms ranged from 1.5 months to 24.5 months, with an average of 15.5 months and standard deviation of 7 months.

The 106 prior mammograms of interval-cancer cases were reviewed independently by the radiologist. The radiologist indicated that 38 of the 106 prior mammographic images had visible architectural distortion, and the remaining 68 images had questionable or no clearly evident architectural distortion; regardless, all of the 106 images were included in the present study. Parts of the images related to or suspected to contain architectural distortion were marked using rectangular boxes based on the reports available on subsequent imaging, biopsy, or by detailed inspection of the prior mammograms. Because each image contained a single site of architectural distortion, one rectangular part per image was marked by the radiologist. The average width, height, and area of the 106 parts of images marked by the radiologist are 56 mm, 39 mm, and 2274 mm^2, with standard deviation of 11.8 mm, 11.6 mm, and 1073.9 mm^2, respectively.

Reports of diagnostic imaging and biopsy were available only for 37 out of the 56 cases of interval cancer. According to the diagnostic imaging or biopsy reports, 27 cases were detected with masses occasionally accompanied by architectural distortion, calcification, or other signs of cancer, and 10 cases were detected with calcification occasionally accompanied by architectural distortion or other signs of cancer. No BI-RADS® ranking [13] of breast density or subtlety of the lesions was provided in the reports from the screening program or the diagnostic clinics.

In addition to the above, all normal cases in the database with at least two visits to the screening program were identified. The mammograms of the penultimate screening visits of the normal cases at the time of preparation of the database were obtained, and labeled as control normal cases. In this manner, 52 mammographic images of 13 normal control cases were obtained for the study.

Figure 4.5 shows two examples of prior mammograms of two interval-cancer cases with the regions of architectural distortion marked with rectangular boxes by the radiologist. Both the prior mammograms of the interval-cancer cases, shown in parts (a) and (b) of Figure 4.5, were acquired 16 months before signs of malignancy were found in the corresponding diagnostic mammograms (not available), and had been declared to be normal at the time of their original acquisition and evaluation.

Figure 1.1 shows a pair of CC and MLO views of a normal breast. Figure 5.1 illustrates another pair of CC and MLO views of a normal control case from the dataset used in the present study; it is seen that the oriented texture in the normal mammograms is coherently aligned.

5.1.2   SCREEN-DETECTED CANCER

In addition to the dataset of interval-cancer cases, all cases of screen-detected cancer with prior mammograms available were included in the second dataset for the present study, with the diagnosis confirmed by biopsy. Mammograms of the affected breasts of seven subjects were obtained. For each subject, the CC and MLO views of the affected breast were used. Mammograms on which cancer was detected were labeled as the *detection mammograms*. The mammograms acquired in the last visit

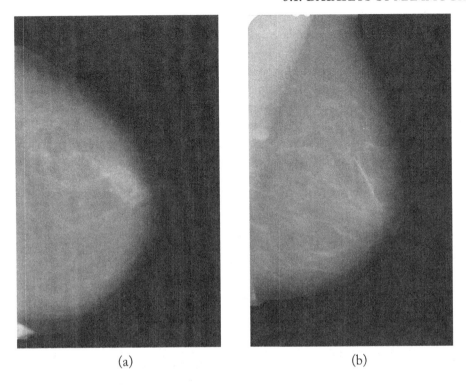

(a)   (b)

Figure 5.1: An example of a normal control case from the dataset used in the present study. (a) CC view; image size 1369 × 850 pixels at 200 μm/pixel. (b) MLO view; image size 1368 × 850 pixels at 200 μm/pixel.

to the screening program prior to the detection of cancer were also included in a separate dataset for the present study, and labeled as *prior mammograms*.

All prior mammograms had been declared to be free of signs of breast cancer at the time of their original acquisition and interpretation in the screening program. The time between the detection and prior mammograms ranged from 12.5 months to 29.2 months, with an average of 23.6 months and standard deviation of 5.2 months. Parts related to architectural distortion were marked by the radiologist in the same manner as described in Section 5.1.1.

The 14 detection mammograms were divided into two categories by the radiologist as containing visible architectural distortion (eight images) and no evident architectural distortion (six images). The average width and height of the parts of architectural distortion in the detection mammograms of screen-detected cases are 40.5 and 26 mm, respectively.

The 14 prior mammographic images were divided into three categories by the radiologist as containing visible architectural distortion (six images), questionable architectural distortion (three images), and no evident architectural distortion (five images). The average width and height of the parts of architectural distortion in the prior mammograms of screen-detected cases are 41 and 26 mm, respectively. Spicules related to architectural distortion were observed to have thickness of the order of 0.8 mm.

Examples of a detection mammogram and a prior mammogram obtained from a case of screen-detected cancer are illustrated in parts (a) and (b) of Figure 5.2, respectively. The parts of architectural distortion marked with rectangular boxes by the radiologist for both the mammograms are also shown. The prior mammogram shown in part (b) was acquired 12.5 months before the detection mammogram shown in part (a), and had been declared to be free of signs of breast cancer at the time of its original acquisition and evaluation.

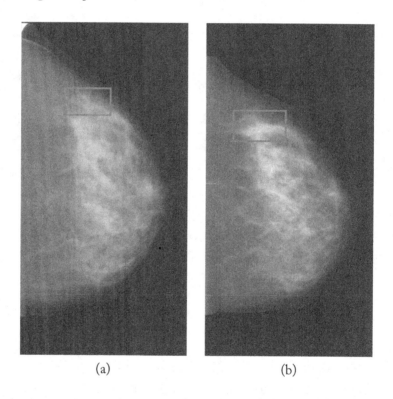

(a)   (b)

Figure 5.2: (a) A "detection mammogram" from a screen-detected cancer case. Image size 1612 × 825 pixels at 200 μm per pixel. The rectangle is of size 45.7 mm × 26.1 mm and indicates the area of architectural distortion identified by the radiologist. (b) The corresponding "prior mammogram" acquired 12.5 months before the detection mammogram in part (a). Image size 1614 × 825 pixels at 200 μm per pixel. The rectangle is of size 51.7 mm × 29.6 mm and indicates the area of architectural distortion identified by the radiologist.

5.2   SELECTION OF ROIS

5.2.1   INTERVAL-CANCER CASES

The image processing methods described in Chapter 4, including the application of Gabor filters and linear phase-portrait analysis, were applied to the 158 images of interval-cancer cases and normal control cases as discussed in Section 5.1.1. From the 158 mammograms in the study, a total of 4,224 ROIs (2,821 ROIs from the 106 prior mammograms of interval-cancer cases with 301 ROIs related to the parts with architectural distortion, and 1,403 ROIs from the 52 normal mammograms) of size 128×128 pixels at 200 μm/pixel (except at the edges of the images) were automatically obtained. The ROIs were labeled at the locations indicated by the peaks in the node maps, in decreasing order of the value of the peak, up to a maximum of 30 ROIs per mammogram. The automatically detected ROIs with their centers within the parts of architectural distortion identified by the radiologist were labeled as TP ROIs; the others were labeled as FP ROIs. Phase-portrait analysis did not detect any TP ROI in one prior mammogram of the interval-cancer cases; the radiologist had indicated that the corresponding image had no evident architectural distortion.

The results of application of the methods described in Chapter 4 and the procedure of selection of ROIs described above are illustrated in Figures 5.3 and 5.4. Parts (a) and (b) of Figure 5.3 show all of the automatically detected ROIs for the prior mammograms of the interval-cancer cases shown in parts (a) and (b) of Figure 4.5, respectively. The parts of architectural distortion, in both the images, are also shown using red rectangular boxes. On the average, the size of the suspicious regions marked on the prior mammograms of interval-cancer cases were observed to be larger than those marked on the detection and prior mammograms of screen-detected cancer cases due to the increased ambiguity and unavailability of the corresponding diagnostic mammograms in the former dataset.

All of the automatically detected ROIs for the normal mammograms shown in parts (a) and (b) of Figure 5.1 are displayed in Figure 5.4.

Table 5.1 gives the details of the dataset of ROIs prepared for further analysis. Figure 5.5 shows some examples of TP and FP ROIs obtained from several cases included in the study.

Table 5.1: Description of the dataset of prior mammograms of interval-cancer and normal cases and the corresponding automatically detected ROIs.

Category	Number of Individuals	Number of Mammo-grams	Number of Detected ROIs	Number of TP ROIs	Number of FP ROIs
Interval cancer	56	106	2821	301	2520
Normal	13	52	1403	0	1403
Total	69	158	4224	301	3923

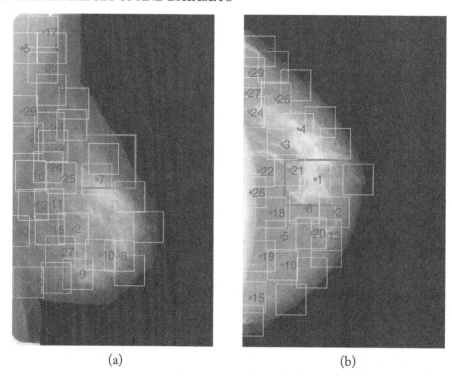

(a)   (b)

Figure 5.3: (a) The 30 ROIs obtained automatically using the peaks detected in the node map for the prior mammogram of an interval-cancer case shown in Figure 4.5 (a). (b) The 29 ROIs obtained automatically using the peaks detected in the node map for the prior mammogram of an interval-cancer case shown in Figure 4.5 (b). See also the related results shown in Figures 4.5, 4.6, 4.7, and 4.8. The size of each ROI is 128 × 128 pixels at 200 μm per pixel (except at the edges). Each asterisk mark (*) corresponds to a peak position detected automatically in the node map. The numbers next to the asterisk marks indicate the peaks in descending order of magnitude. In each image, the ROI automatically detected and labeled first in the node map analysis is within the part of architectural distortion.

With the ROIs obtained automatically as above, the highest sensitivity of detection of architectural distortion was determined to be 0.99 at the FP rate of 24.8 per image for the set of interval-cancer cases including normal control cases.

5.2.2   SCREEN-DETECTED CANCER CASES

With the second dataset of screen-detected cancer cases including prior and detection mammograms, phase-portrait analysis did not detect any site of architectural distortion in three of the prior mammograms and one of the detection mammograms; the radiologist had indicated that the cor-

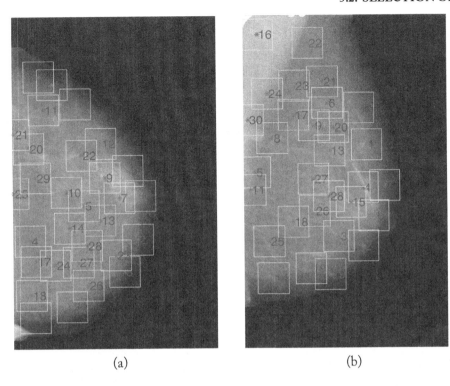

(a)   (b)

Figure 5.4: The 30 ROIs obtained automatically using the peaks detected in the node map for the mammograms of a normal control case shown in (a) Figure 5.1 (a), and (b) Figure 5.1 (b).

responding images had no clearly evident architectural distortion. No TP ROI was obtained for these four images; therefore, the maximum sensitivity achievable was 0.79 and 0.93 with the prior and detection mammograms, respectively. From the 28 mammograms in the second dataset, a total of 784 ROIs of size 128×128 pixels (except at the edges of the images) at 200 μm/pixel were automatically obtained; Table 5.2 summarizes this dataset.

Table 5.2: Description of the dataset with 14 prior mammograms and 14 detection mammograms from seven screen-detected cancer cases.

Category	Number of Images	Number of Detected ROIs	Number of TP ROIs	Number of FP ROIs
Detection mammograms	14	398	18	380
Prior mammograms	14	386	21	365
Total	28	784	39	745

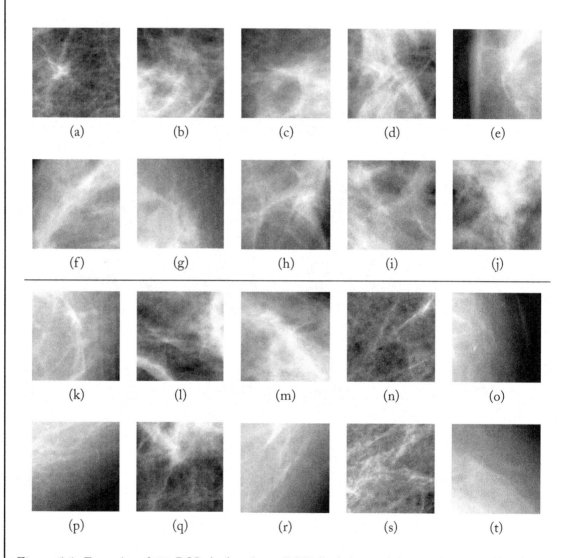

Figure 5.5: Examples of TP ROIs (a–j) and FP ROIs (k–t) detected. Each ROI is of size 128×128 pixels at 200 μm/pixel resolution, or 25.6×25.6 mm. Reproduced with permission from: R. M. Rangay-yan, S. Banik, and J. E. L. Desautels. "Computer-aided Detection of Architectural Distortion in Prior Mammograms of Interval Cancer." *Journal of Digital Imaging*, vol. 23(5), pp. 611–631, October 2010. © Springer, SIIM.

With the ROIs obtained automatically as above, the highest sensitivity of detection of architectural distortion was determined to be 0.93 at the FP rate of 27.1 per image for the set of detection mammograms and 0.79 at the FP rate of 26.1 per image for the set of prior mammograms.

Parts (a) and (b) of Figure 5.6 show all of the automatically detected ROIs for the detection mammogram and the prior mammogram of the screen-detected cancer case displayed in Figure 5.2, respectively.

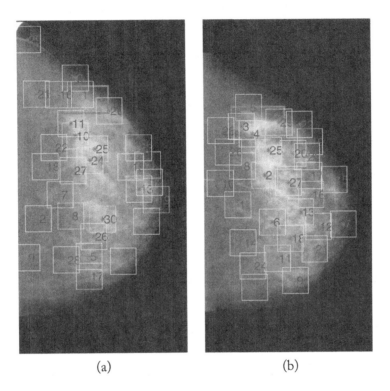

(a)   (b)

Figure 5.6: (a) The 30 ROIs obtained automatically using the peaks detected in the node map for the detection mammogram shown in Figure 5.2 (a). The size of each ROI is 128×128 pixels at 200 μm per pixel (except at the edges). Each asterisk mark (*) corresponds to a peak position detected automatically in the node map. The numbers next to the asterisk marks indicate the peaks in descending order of magnitude. (b) The ROIs obtained automatically using the peaks detected in the node map for the corresponding prior mammogram shown in Figure 5.2 (b).

5.3   REMARKS

In this chapter, the datasets used in the present work and the experimental set up were described. The main dataset includes prior mammograms of interval-cancer cases and normal control cases.

The second dataset includes detection and prior mammograms of screen-detected cancer cases; this dataset is used for comparison and cross-validation purposes. Several techniques for image processing, feature extraction, analysis of classification performance, feature selection, and pattern recognition were developed and deployed for the detection of architectural distortion in prior mammograms of interval-cancer cases, and are described in the following chapters.

CHAPTER 6

Feature Selection and Pattern Classification

6.1 ANALYSIS OF FEATURES AND PATTERN CLASSIFICATION

The final step of medical image analysis is to perform classification of the given image or ROI based on the features that have been extracted from the image. For accurate classification, the features extracted or detected should be able to represent each of the categories efficiently. However, in practical applications, most of the features may overlap in the ranges of their values and provide inadequate discrimination between multiple classes; consequently, a number of parameters or features may be required for accurate classification.

In extracting or designing features, one would like to find features that are simple to extract, invariant to transformation of the original image, insensitive to noise, and useful for discriminating patterns in different categories [139]. The choice of the distinguishing features is a critical step and depends on the characteristics and the prior knowledge of the problem. However, incorporating prior knowledge could be subtle and difficult. In this context, analysis of the separability and the robustness of the extracted features is an important part of feature selection and pattern classification.

Feature selection is an essential step before classification. In practical applications, use of all of the features available could be computationally expensive due to the large dimension of the feature space, or a large number of features with a limited amount of observations could make the classification process unreliable and incapable of generalization [140, 141]. In addition, irrelevant or redundant features may affect negatively the accuracy of classification algorithms and increase the storage requirements. Reducing the number of features may help in making the classification models easier to implement and understand. Feature selection [139, 141, 142] can be performed by evaluating the performance of each feature or combinations of several features based on their p-values [143], AUC values [17, 18], statistical or distance measures, deviation or error measures, discriminant analysis, genetic programming or algorithms [141], classification accuracy using a training set [142, 144], or through exhaustive search. This chapter presents descriptions of the methods used in the present work for feature extraction, feature selection, and pattern recognition.

6.2 CHARACTERIZATION OF FEATURES

The conceptual boundary between feature extraction and classification is arbitrary [139]. An ideal feature extractor should produce a form or representation of the features that makes the classification task simple; on the contrary, a perfect classifier should not need to use a sophisticated feature extractor. A typical feature extractor characterizes an object to be recognized by producing similar values for objects in the same category and significantly different values for objects in different categories [139]. In the case of supervised classification, where some or all of the samples are labeled, the p-value obtained via Student's t-test or the A_z values obtained via ROC analysis could be used to characterize or quantify the discrimination power and reliability of each feature. In an ideal scenario, any feature with a high A_z value or a low p-value is likely to yield good discrimination between the classes. However, in real life, different methods of analysis of the separability of features may not concur; a combination of features with high levels of performance selected using a particular method of analysis might not result in the best possible classification performance. Despite many criticisms, the p-value and the A_z value are the most popular criteria for characterization of the separability or performance of individual features, especially in the areas of medicine, biology, and engineering research [143].

In the present work, the p-value and the A_z value are used to assess the performance of a particular feature; however, they are not directly used in the process of feature selection or pattern classification. The classification performance of a combination of particular types or sets of features or combinations of features selected with a feature selection method is analyzed using ROC and FROC analysis.

6.2.1 STUDENT'S t-TEST AND THE p-VALUE

Student's t-test is a conventional statistical procedure for measuring the significance of a difference of means [143, 145]. After performing the t-test, a t value is determined and can be used to obtain a p-value by using the table of values from Student's t distribution. The p-value is the probability that the test statistic t is at least as extreme as the one that was actually observed, assuming that the null hypothesis were true. A null hypothesis is a statistical hypothesis that is tested for possible rejection under the assumption that it is true and typically corresponds to a general or default position. A smaller p-value indicates a stronger evidence against the null hypothesis than a larger value; when the null hypothesis is rejected, the result is considered to be statistically significant.

When two distributions are considered to have the same variance, but possibly different means, the Student's t parameter can be computed with the null hypothesis being that the means are equal. The standard error of the difference of the means s_D is given by [145]

$$s_D = \sqrt{\frac{\sum_{x_i \in C_1}(x_i - m_1)^2 + \sum_{x_i \in C_2}(x_i - m_2)^2}{N_1 + N_2 - 2}\left(\frac{1}{N_1} + \frac{1}{N_2}\right)}, \qquad (6.1)$$

where each summation is taken over all the points in one set of samples, the first or second; m_1 and m_2 refer to the means of the first and second sets of samples; C_1 and C_2 are two classes of the samples x_i; and N_1 and N_2 are the numbers of points in the first and second sets of samples, respectively. Then, the variable t is computed as

$$t = \frac{m_1 - m_2}{s_D}. \tag{6.2}$$

The significance of the value of t for Student's distribution is evaluated with $N_1 + N_2 - 2$ degrees of freedom.

If two distributions are considered to have largely different values of the variance, the appropriate statistic for the unequal variance t-test is given by [145]

$$t = \frac{m_1 - m_2}{[Var(X_1)/N_1 + Var(X_2)/N_2]^{1/2}}, \tag{6.3}$$

where X_1 and X_2 represent the two subsets of samples corresponding to classes C_1 and C_2, respectively, and Var represents the variance. The statistic is distributed approximately as the Student's t distribution with a number of degrees of freedom given by [145]

$$\frac{\left[\frac{Var(X_1)}{N_1} + \frac{Var(X_2)}{N_2}\right]^2}{\frac{[Var(X_1)/N_1]^2}{N_1-1} + \frac{[Var(X_2)/N_2]^2}{N_2-1}}. \tag{6.4}$$

In the case of paired samples, the Student's t test formulas are defined as follows [145]:

$$Cov(X_1, X_2) \equiv \frac{1}{N-1} \sum_{i=1}^{N} (x_{1i} - m_1)(x_{2i} - m_2), \tag{6.5}$$

$$s_D = \left[\frac{Var(X_1) + Var(X_2) - 2\,Cov\,(X_1, X_2)}{N}\right]^{1/2}, \tag{6.6}$$

$$t = \frac{m_1 - m_2}{s_D}, \tag{6.7}$$

where x_{ki} represents the ith sample in the kth class, Cov represents the covariance, and N is the number of pairs of samples. The significance of the t statistic in Equation 6.7 is evaluated for $N - 1$ degrees of freedom.

A typical approach is to indicate only that the p-value is smaller than 0.05 ($p < 0.05$) or smaller than 0.01 ($p < 0.01$). When the p-value is between 0.05 and 0.01, the result is considered "statistically significant"; when it is less than 0.01, the result is considered to be "highly statistically significant" [143]. In the present work, the statistical significance (i.e., the p-value) of differences between the ROC curves was assessed using ROCKIT [146]. Other p-values presented in the present work, except the significance of the differences between the FROC curves, were obtained by using the two-tailed t-test [143] in MATLAB® [147].

6.2.2   THE RECEIVER OPERATING CHARACTERISTIC (ROC) CURVE

In medical imaging, ROC analysis [17, 18] is a frequently used procedure for evaluation of the performance of a radiologist or a CAD system in the detection of abnormalities in medical images. In conventional ROC methodology, the sensitivity and specificity of a diagnostic procedure are considered; the sensitivity and specificity values of the procedure generally depend on the particular decision criteria adopted by the user to distinguish the nominally positive cases (e.g., images, ROIs, or features) from nominally negative cases. A decision method or criterion is considered to be good if it simultaneously has a high sensitivity and a high specificity. Both the sensitivity and specificity are usually presented in percentages or fractions; see Section 1.3 for details. The ROC curve illustrates the sensitivity versus the FPR for various values of the decision threshold and is a useful analysis tool for binary classification problems. The AUC or A_z could be used as a performance criterion.

Let Ω denote the set of elements being classified (e.g., mammographic ROIs), and $\omega \in \Omega$ be an element in the set. A classifier can be comprised of a function $g : \Omega \mapsto \mathbb{R}$ and a decision threshold T such that if $g(\omega) > T$, the element ω is classified as positive; otherwise, ω is classified as negative. The value $g(\omega)$ is referred to as the discriminant value of ω [17, 18].

In generating an ROC curve, the sensitivity and specificity of the classifier depend on the threshold value T. Typically, a high value of T produces a strict classifier allowing only a few negative elements to be classified as positives; however, many positive elements may not be able to overcome the high threshold value as well. Consequently, the corresponding classifier will have high specificity but low sensitivity. On the contrary, a low value of T will allow the classifier to accept more elements as positives; in this case, the misclassification rate will increase, resulting in high sensitivity but low specificity. For this reason, an ROC curve can be regarded as an effective tool for visualizing, analyzing, and selecting classifiers based on their performance [148].

In the present work, ROC curves, AUC or A_z values, and the associated p-values were obtained by using ROCKIT, a widely used software package developed at the University of Chicago, IL [146]. ROCKIT uses maximum likelihood estimation to fit a binormal ROC curve to continuously distributed data or ordinal category data. ROCKIT also calculates the statistical significance of differences between ROC index estimates and parameters. On the basis of a "bivariate binormal" model, ROCKIT allows for the comparison of two paired, partially paired, or unpaired datasets with regard to differences in the binormal ROC curve parameters a and b (related to the difference in the mean and of the values of the standard deviation of the two latest normal distributions used in the fit), difference in the areas (A_z) under the two estimated binormal ROC curves, and difference between the two TPRs on the two curves at a selected FPR.

6.2.3   FROC ANALYSIS

Conventional ROC methodology is applicable in the binary classification problem only and cannot deal with more than two decision alternatives; each case is required to be assigned to one of two defined categories [149]. In addition, ROC techniques account for only a single decision per case and cannot allow multiple reports for different lesions of the same type (e.g., malignant tumors); the method cannot make full utilization of the potentially available diagnostic information obtained from multiple observers, such as the diagnostic classification or location of detected lesions. Researchers have worked on modifications to the ROC methodology to extend it to encompass conditions frequently reported in clinical studies, such as indications of lesion location and/or the presence of multiple lesions [149].

FROC is a method of collecting observer performance data where the observer marks and rates suspicious regions in the images [21, 150, 151]. The record of mark-rating pairs constitutes the FROC that refers to a method of plotting free-response data as a set of operating points; the FROC curve is a theoretical fit to the FROC operating points.

In the present work, FROC analysis was used to assess the FP rate for a given level of sensitivity when the classification of automatically detected and segmented ROIs was placed in the context of detection of architectural distortion in full mammograms. Because each image of interval-cancer cases has only one site of architectural distortion, only one TP ROI with the highest discriminant value per image was considered when generating the FROC curves, and additional TP ROIs (if any) in the same image were discarded. Because the dataset of interval-cancer cases includes 100 paired MLO and CC views from 50 patients (one affected breast per patient), six individual images (either MLO or CC view as the radiologist could not find any suspicious region in the other view) from six patients, and 52 normal images from 13 patients (CC and MLO views for both breasts), only the leave-one-image-out method is applied for consistent reporting in FROC analysis.

The jackknife alternative FROC (JAFROC) software package [21, 150, 151] was used for statistical analysis of FROC data. JAFROC compares the performance of one or more readers interpreting the same set of images or cases in two or more modalities. Readers could be radiologists, mammographers, or computer algorithms designed to find lesions, such as CAD methods. JAFROC can be used to evaluate whether or not CAD improves the diagnostic performance of a radiologist, or if one CAD system is better than another. JAFROC is a nonparametric method of quantifying search performance not involving FROC curves and cannot be used to predict such curves [150]. JAFROC was used in the present work to evaluate the statistical significance (i.e., the p-value) of the differences between FROC curves using the jackknife method. Jackknife is a resampling technique in which cases are successively removed or jackknifed, the figure-of-merit is recomputed, and a pseudovalue is calculated and analyzed using an analysis of variance technique [150].

6.3   FEATURE SELECTION

In performing classification tasks, numerous features may be extracted, but only a small number of them may have high discriminative capability [140]. The removal of noisy and irrelevant features can help to improve the performance of the classifier. In addition, extensive amount of time, labor, and expense can be saved by only extracting high-performance features. Nevertheless, the following problems make feature selection a challenging task [140].

- **Linearly nonseparable classes:** This is a common phenomenon in real-world applications. In image classification, the semantic gap between the high-level concepts used by humans to interpret visual content and the low-level features used by computers for classification usually results in a nonlinear relationship between them [140].

- **Quick response:** A fast response in the training process is always desirable, especially in the case of real-time processing. For this reason, a computationally efficient criterion is usually preferred; such a criterion could be combined with a more sophisticated and computationally more expensive selection technique for further improvement in the selection and classification performance. However, in a real-world application, when a CAD technique is used to analyze an unknown new case, the computational efficiency of the feature selection technique may not be important.

- **Selection with a small number of samples:** If the extraction of a large number of features from all the samples becomes expensive, complex, or time consuming, prior identification of the most useful features with a small-sized sample set could be useful. Consequently, attention may be given to the extraction of only the most useful features in the future. However, this approach would require the selection criterion to be not sensitive to the sampling procedure [140].

- **Noisy features:** Some features that are statistically irrelevant to class labels or are corrupted by noise during data generation could create ambiguity and complexity in the classification process. In this context, a robust selection criterion would be needed.

Feature selection, more precisely feature subset selection, is the process of finding a reduced set of r features out of the given set of d features according to a given selection criterion [140]. In a classification-oriented feature selection procedure, the r selected features are expected to produce lower classification errors. Feature selection usually comprises a selection criterion and a search strategy; because feature selection is essentially a combinatorial optimization problem, the choice of an efficient search strategy is important [140]. Many search algorithms have been reported in the literature, including the sequential forward selection, sequential backward selection, stepwise selection, and floating search methods [152].

Sequential Forward Selection

In sequential forward selection, starting with an empty set of candidate features, features are sequentially added to the set until the addition of further features does not improve the classification performance [147]. At first, the best single feature in terms of separation of the classes of interest (e.g., p-value) is determined and added to the empty set of selected features. The selected feature is then used in combination with each of the other features, in turn, to create pairs; the performance of each pair is evaluated and the best pair is selected for the next iteration. Subsequent features are selected in the same manner and the process continues until the required number of features is selected, or the classification performance does not improve significantly with the inclusion of additional features [141, 153]. Sequential forward selection usually includes the better features in the best (final) set, and is frequently used in feature selection and pattern classification for its computational efficiency as compared to the exhaustive search method [141].

Sequential Backward Selection

Starting with the complete set of given features, sequential backward selection eliminates each feature, in turn, from the set until the removal of further features deteriorates the classification performance. At first, the (worst) feature, the elimination of which causes the largest separation of the classes of interest in the remaining feature space, is identified and removed permanently. The process continues in the same manner by removing features one by one, until the desired number of features remain, or the removal of any more features causes a deterioration in the classification performance.

Stepwise Selection

This approach is a combination of the sequential forward and backward selection methods, in which tests are performed at each stage to determine which feature should be included or excluded.

Floating Search Method

Floating search methods are sequential search methods characterized by a dynamically changing number of features to be included or eliminated at each step. Such methods have been shown to give good results and to be computationally efficient [154].

6.3.1   LOGISTIC REGRESSION

Logistic classification is a widely used statistical method for feature selection and/or classification based on the probability of occurrence of an event by fitting data to a logistic regression model [12, 153]. The logistic regression model is a generalized linear model used for binomial regression and is applicable to problems associated with the classification of patterns into one of two classes.

Studies indicate that in the case of a binary response variable, the logistic response function is usually curvilinear, and is referred to as a sigmoidal function with a typical shape of either a tilted "S" or as a reverse tilted "S." The function can be considered linear except at the shoulder or toe region with asymptotes at 0 and 1.

In pattern classification using logistic regression, an event is typically defined by the membership of a pattern vector in one of the two classes under consideration [12]; based on the given parameters, a response variable constrained to the range [0, 1] is computed. As a result, the variable may be considered as the probability of belonging to a class, and the probability of the pattern vector belonging to the other class can be obtained by the calculating the difference between unity and the estimated value for the former class.

In the case of an M-dimensional feature vector $\mathbf{x}$, the model is given by

$$P(event) = \frac{\exp(z)}{1 + \exp(z)},$$

(6.8)

or equivalently,

$$P(event) = \frac{1}{1 + \exp(-z)},$$

(6.9)

where z is the linear combination of the features x_k, with $k = 1, 2, \ldots, M$, and is given by

$$z = w_0 + w_1 x_1 + w_2 x_2 + \ldots + w_M x_M = \mathbf{w}^T \mathbf{x}_a,$$

(6.10)

that is, z is the dot product of the augmented feature vector $\mathbf{x}_a = [1, x_1, x_2, \ldots, x_M]^T$ with a coefficient vector or weight vector $\mathbf{w} = [w_0, w_1, \ldots, w_M]^T$.

The classification of a given sample $\mathbf{x}$ in a two-class problem is performed as

$$\mathbf{x} \in \begin{cases} C_1 & \text{if } P(event) < T \\ C_2 & \text{otherwise} \end{cases},$$

(6.11)

where T is a threshold.

The common approaches for implementation of logistic regression include forward selection, backward selection, and stepwise selection. In stepwise logistic regression, the determination of the predictive variables is performed by an automatic and iterative procedure [153]: each step consists of one step of forward selection and one step of backward elimination; the process is continued until no features can be added or removed.

6.3.2   STEPWISE REGRESSION

Stepwise regression is a systematic statistical method for feature (or model) selection: features are added or removed from a multilinear model based on their statistical significance in a regression [155]. Feature selection through stepwise regression iteratively varies the number of features by adding features into or removing features from the group of selected features based on a selection criterion using F-statistics [142].

Stepwise regression is an automatic model selection procedure and can be applied to the situation with a large number of potential explanatory variables but lacks any specific criteria to depend on for selection of the model. The method is generally used in regression analysis; however, the approach can be applicable to different forms of model selection. Stepwise regression can be considered as a variant of forward selection: at each stage, a new variable (feature) is added, and the obtained model is checked for any possible elimination of the selected features without significantly increasing the residual error. The method continues until the error can be considered to have been (locally) minimized, or the available improvement goes below some critical value.

The method starts with an initial model and then evaluates the discriminating or explanatory power of incrementally larger and smaller models at each iteration [155]. At each iteration, the p-value of an F-statistic is computed to evaluate the models with and without a potential term; if the term (feature) is not in the existing model, the null hypothesis becomes that the term will have a zero coefficient if added to the model. The term can be added to the model if sufficient evidence can be found to reject the null hypothesis. On the other hand, for an existing term in the model, the null hypothesis becomes that the term has a zero coefficient. If sufficient evidence cannot be found to reject the null hypothesis, the term is removed from the model. The method proceeds as follows [155, 156]:

1. Start with an arbitrary initial model (or set of selected features).

2. If any term is currently not in the model and the term has a p-value less than a specified entrance tolerance, add the term with the smallest p-value in the model, and repeat this step; otherwise, go to the next step.

3. If any term in the model has a p-value greater than a specified exit tolerance, remove the term with the highest p-value and go to Step 2; otherwise, stop.

Depending on the terms included in the initial model and the order in which terms are moved in and out of the model, the method may generate different models from the same set of potential terms. The method stops when no single step can significantly improve the model. However, one cannot provide any assurance that a different initial model or a different sequence of steps will not lead to a better fit. In this sense, stepwise models are not unique and are locally optimal, but may not be globally optimal [156].

6.4   PATTERN CLASSIFICATION

In machine learning, pattern recognition deals with the assignment of a label to a given input according to a specific algorithm [12]. A variant of pattern recognition is pattern classification, which attempts to assign each input value to one of a given set of classes. Algorithms for pattern recognition depend on the type of label output, the type of learning (i.e., supervised or unsupervised), and the nature of algorithm (i.e., statistical or deterministic). Binary classification is the task of classifying the input samples into two groups on the basis of whether they have some common property or not, such as being normal or abnormal in medical diagnosis. Multiclass classification is the task of assigning the input samples into one of the multiple categories under consideration. In this section, some of the widely used tools for binary pattern classification are described.

6.4.1   FISHER LINEAR DISCRIMINANT ANALYSIS

Fisher linear discriminant analysis (FLDA) is a linear classification technique based on a linear projection of the given M-dimensional feature data, $\mathbf{x}_i$, onto a line, with the expectation that such projections onto a line will be well separated by class [12]. Thus, the line is oriented in a direction that maximizes the class separation. FLDA is useful when the size of the feature vector is large. For a two-class example [157], the given set,

$$X = \{\mathbf{x}_1, \mathbf{x}_2, \ldots, \mathbf{x}_N\} = \{X_1, X_2\}, \tag{6.12}$$

is partitioned into $N_1 \leq N$ training vectors in subset X_1, corresponding to class C_1, and $N_2 \leq N$ training vectors in set X_2, corresponding to class C_2, where $N_1 + N_2 = N$. The projections of the feature vectors can be formed via

$$y_i = \mathbf{w}^T \mathbf{x}_i = \langle \mathbf{w}, \mathbf{x}_i \rangle, \quad i = 1, 2, \ldots, N, \tag{6.13}$$

where $\mathbf{w} = [w_1, w_2, \ldots, w_M]^T$ is a weight vector that represents the projection operator, and y_i is the resulting discriminant value. If a further constraint is imposed as $\|\mathbf{w}\| = 1$, each y_i can be regarded as the projection of $\mathbf{x}_i$ onto a line in the direction of $\mathbf{w}$. The problem is to find the direction of $\mathbf{w}$, given X, such that y_i from X_1 and y_i from X_2 ideally fall into distinct clusters along the line, denoted by Y_1 and Y_2.

A measure of the separation of the projections can be given by the difference of the means of the projections; for example, $|\mu_{Y1} - \mu_{Y2}|^2$, where μ represents the mean, can be considered as a measure of separation of the projections. This measure is related to the sample means of X_1 and X_2 through $\mathbf{w}$ as

$$\mathbf{m}_k = \frac{1}{N_k} \sum_{\mathbf{x}_i \in X_k} \mathbf{x}_i, \tag{6.14}$$

where $\mathbf{m}_k$ is the sample mean of the vectors in X_k.

Thus, the mean of the projections for each class is a scalar quantity, given as

$$\begin{aligned} \overline{m}_k &= \frac{1}{N_k} \sum_{y_i \in Y_k} y_i = \frac{1}{N_k} \sum_{\mathbf{x}_i \in X_k} \mathbf{w}^T \mathbf{x}_i \\ &= \mathbf{w}^T \frac{1}{N_k} \sum_{\mathbf{x}_i \in X_k} \mathbf{x}_i = \mathbf{w}^T \mathbf{m}_k. \end{aligned} \tag{6.15}$$

Consequently, the difference between the means of the projections using the sample data for two classes is

$$|\overline{m}_1 - \overline{m}_2| = |\mathbf{w}^T (\mathbf{m}_1 - \mathbf{m}_2)|. \tag{6.16}$$

A good classifier cannot be characterized by the difference between the means of the projected data alone; the variance of y_i in Y_k relative to the means also needs to be considered. Therefore, a better measure of class separation can be obtained by considering the means and variances of the classes of concern, and can be given by the ratio of the difference of the means to the sum of the variances of the within-class data. For example, a reasonable criterion for a two-class case can be given by

$$J(\mathbf{w}) = \frac{(\mu_{Y1} - \mu_{Y2})^2}{\sigma_{Y1}^2 + \sigma_{Y2}^2}, \tag{6.17}$$

or, considering the case of sample data,

$$J(\mathbf{w}) = \frac{(\overline{m}_1 - \overline{m}_2)^2}{\hat{\sigma}_1^2 + \hat{\sigma}_2^2}, \tag{6.18}$$

where $\hat{\sigma}_k^2$ is a measure of the within-class scatter of the projected data.

The within-class scatter of the projected data can be defined as

$$\overline{s}_k^2 = \sum_{y_i \in Y_k} (y_i - \overline{m}_k)^2. \tag{6.19}$$

In this case, $\hat{\sigma}_k^2 = \frac{1}{N_k - 1} \sum_{y_i \in Y_k} (y_i - \overline{m}_k)^2$ is an unbiased estimator of the variance. Consequently, for the N_k samples in X_k,

$$\overline{s}_k^2 \approx (N_k - 1)\hat{\sigma}_k^2. \tag{6.20}$$

Fisher (see [157]) showed that a reasonable measure of separability of the projected data is the criterion function that is a scaled version of Equation 6.18:

$$J(\mathbf{w}) = \frac{(\overline{m}_1 - \overline{m}_2)^2}{\overline{s}_1^2 + \overline{s}_2^2}. \tag{6.21}$$

This is an optimization problem to determine the direction of $\mathbf{w}$ such that the measure in Equation 6.21 is at its maximum. The vector $\mathbf{w}$ that maximizes Equation 6.17, 6.18, or 6.21 is used in a linear discriminant function, given in the form $\mathbf{w}^T \mathbf{x}$, to generate the Fisher linear discriminant function.

Then, classification of a given sample $\mathbf{x}$ in a two-class problem can be performed by

$$\mathbf{x} \in \begin{cases} C_1 & \text{if } y = \mathbf{w}^T \mathbf{x} < T \\ C_2 & \text{otherwise} \end{cases}, \tag{6.22}$$

where T is a threshold.

6.4.2   THE BAYESIAN CLASSIFIER

Bayesian decision theory, a fundamental statistics-based approach, is widely used in pattern recognition [139]. The approach assumes that the decision problem can be represented in probabilistic terms and that all the relevant prior probabilities are known; the method quantifies the tradeoffs between various classification decisions using probability and the costs associated with such decisions [139].

Let $\{C_1, C_2, \ldots, C_J\}$ be the finite set of J states or categories, and let $\{\alpha_1, \alpha_2, \ldots, \alpha_K\}$ be the finite set of K possible actions. The cost (loss) function $\lambda(\alpha_i | C_j)$ gives the cost associated for taking the action α_i when the state is C_j.

Let the feature vector $\mathbf{x}$ be an M-component vector of random variables with the associated (unconditional) probability density function (PDF) $p(\mathbf{x})$, and let $p(\mathbf{x}|C_j)$ be the state-conditional PDF for $\mathbf{x}$, with the PDF for $\mathbf{x}$ conditioned on C_j being the true state. $P(C_j)$ is the prior probability of state C_j. Then, the posterior probability $P(C_j|\mathbf{x})$ can be computed from $p(\mathbf{x}|C_j)$ by using Bayes' formula as [139]

$$P(C_j|\mathbf{x}) = \frac{p(\mathbf{x}|C_j)P(C_j)}{p(\mathbf{x})}, \tag{6.23}$$

where

$$p(\mathbf{x}) = \sum_{j=1}^{J} p(\mathbf{x}|C_j)P(C_j). \tag{6.24}$$

Now, if a particular $\mathbf{x}$ is observed and an action α_i is taken, $\lambda(\alpha_i|C_j)$ represents the associated cost for the true state being C_j. Because $P(C_j|\mathbf{x})$ is the probability that the true state is C_j, the expected cost associated with taking the action α_i can be given by

$$R(\alpha_i|\mathbf{x}) = \sum_{j=1}^{J} \lambda(\alpha_i|C_j)P(C_j|\mathbf{x}). \tag{6.25}$$

Using the decision-theoretic terminology [139], an expected cost (loss) for taking an action can be referred to as a *risk*, and $R(\alpha_i|\mathbf{x})$ can be regarded as the *conditional risk*. Whenever any particular $\mathbf{x}$ is observed, the expected cost can be minimized by selecting the action that minimizes the conditional risk. Therefore, the optimization problem is to find a decision rule that minimizes the overall risk R (the expected cost associated with a given decision rule).

Because $R(\alpha_i|\mathbf{x})$ is the conditional risk associated with action α_i, and because the decision rule specifies the action, the overall risk can be given as [139]

$$R(\alpha_i) = \int R(\alpha_i|\mathbf{x})p(\mathbf{x})d\mathbf{x}, \tag{6.26}$$

where $d\mathbf{x}$ is the notation for an M-dimensional volume element, and the integral extends over the entire feature space. Now, if α_i is chosen so that $R(\alpha_i)$ is as small as possible for every $\mathbf{x}$, the overall risk of taking the action α_i will be minimized. The resulting minimum overall risk is referred to as *Bayes risk*, and represents the best performance that can be achieved.

For a two-category classification problem, let the action α_1 correspond to deciding that the true state is C_1, and action α_2 correspond to deciding that it is C_2. Let $\lambda_{ij} = \lambda(\alpha_i|C_j)$ be the cost associated with deciding C_i when the true state is C_j. Using the conditional risk given by Equation 6.25, it can be shown that [139]

$$\begin{aligned}
R(\alpha_1|\mathbf{x}) &= \lambda_{11}P(C_1|\mathbf{x}) + \lambda_{12}P(C_2|\mathbf{x}), \\
R(\alpha_2|\mathbf{x}) &= \lambda_{21}P(C_1|\mathbf{x}) + \lambda_{22}P(C_2|\mathbf{x}).
\end{aligned} \tag{6.27}$$

A number of ways can be found for expressing the minimum-risk decision rule; each has its own minor advantages [139]. The fundamental rule is to decide on or select C_1 if $R(\alpha_1|\mathbf{x}) < R(\alpha_2|\mathbf{x})$. In terms of the posterior probabilities, C_1 is selected if

$$(\lambda_{21} - \lambda_{11})P(C_1|\mathbf{x}) > (\lambda_{12} - \lambda_{22})P(C_2|\mathbf{x}). \tag{6.28}$$

Using Bayes' formula, the posterior probabilities can be replaced by the prior probabilities and the conditional densities, resulting in an equivalent rule to select C_1 if

$$(\lambda_{21} - \lambda_{11})p(\mathbf{x}|C_1)P(C_1) > (\lambda_{12} - \lambda_{22})p(\mathbf{x}|C_2)P(C_2), \tag{6.29}$$

or C_2 otherwise.

An alternative, which follows under the assumption that $\lambda_{12} > \lambda_{22}$ and $\lambda_{21} > \lambda_{11}$, is to select C_1 if

$$\frac{p(\mathbf{x}|C_1)}{p(\mathbf{x}|C_2)} > \frac{\lambda_{12} - \lambda_{22}}{\lambda_{21} - \lambda_{11}} \frac{P(C_2)}{P(C_1)}. \tag{6.30}$$

This form of the decision rule is focused on the dependence of the PDFs on $\mathbf{x}$. One could consider $p(\mathbf{x}|C_j)$ to be a function of C_j and then form the likelihood ratio $p(\mathbf{x}|C_1)/p(\mathbf{x}|C_2)$. Thus, Bayes' decision rule can be interpreted as follows: select C_1 if the likelihood ratio is larger than a threshold value which is independent of the observation $\mathbf{x}$ [139].

For a binary classification problem, consider a set of observations $\mathbf{x}$, with $p(\mathbf{x}|C_1)$ and $p(\mathbf{x}|C_2)$ being both normally distributed with mean and covariance $(\boldsymbol{\mu}_1, \mathbf{S}_1)$ and $(\boldsymbol{\mu}_2, \mathbf{S}_2)$, respectively. Under the assumption stated above, Bayes' optimal solution can be obtained by classifying the component as being from C_1 if the log-likelihood ratio is less than some threshold T [139, 157], as follows:

$$(\mathbf{x} - \boldsymbol{\mu}_2)^T\mathbf{S}_2^{-1}(\mathbf{x} - \boldsymbol{\mu}_2) + \ln|\mathbf{S}_2| - (\mathbf{x} - \boldsymbol{\mu}_1)^T\mathbf{S}_1^{-1}(\mathbf{x} - \boldsymbol{\mu}_1) - \ln|\mathbf{S}_1| + \ln\frac{P(C_1)}{P(C_2)} < T \tag{6.31}$$

where T is independent of the observation $\mathbf{x}$.

Without any further assumption, the classification procedure as above is typically referred to as quadratic discriminant analysis (QDA). Although the classifier as above does not consider the state-conditional distributions of the independent features related to the formulation of the commonly known "Naïve Bayes Classifier," it is referred to as the Bayesian classifier in this book.

6.4.3   NEURAL NETWORKS

In practical applications of pattern recognition, prior knowledge about the probabilities of an element or pattern belonging to the classes of concern could be unavailable; classical methods of pattern classification may not be applicable in such cases [139]. In this context, a classifier that can acquire and store relevant knowledge from its environment through a learning process would be more suitable. ANNs are such classifiers and could be effective in solving complex and nonlinear classification problems through adaptive learning, input-output mapping, and evidential response [158].

The development of ANNs is motivated by the recognition that the human brain works in a different way from a conventional computer algorithm. Similar to the human brain's structure, a neuron is the fundamental basic unit for information processing in an ANN. Figure 6.1 shows the model of a neuron that forms the basis for designing ANNs [158]. In a neuron, three basic elements can be identified:

- A set of synapses or connecting links, each of which is characterized by a weight or strength of its own.

- An adder for performing summation of the input signals, weighted by the respective synapses of the neuron. An adder in the neuron works as a linear combiner.

- An activation function for restricting the amplitude of the output of a neuron to some finite range of values.

A typical neuronal model, as shown in Figure 6.1, also includes an externally applied bias, denoted by b_k. Depending on the linearly added input value being positive or negative, b_k increases or lowers the net input to the activation function, respectively.

In mathematical terms, a neuron can be described using the following set of equations [158]:

$$u_k = \sum_{j=1}^{M} w_{kj} x_j,$$
$$y_k = \varphi(u_k + b_k), \tag{6.32}$$

where $x_1, x_2, \ldots, x_M$ are the input values; $w_{k1}, w_{k2}, \ldots, w_{kM}$ are the synaptic weights of neuron k; u_k is the output of the linear combiner; $\varphi(.)$ is the activation function; and y_k is the output of the neuron.

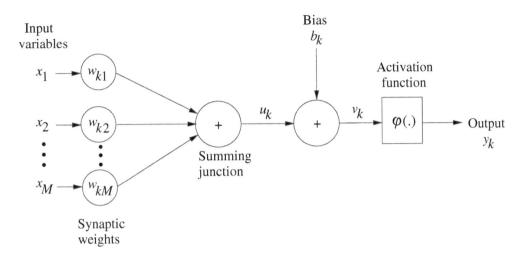

Figure 6.1: Nonlinear model of a neuron.

An affine transformation is applied to the output u_k of the linear combiner by the bias b_k, as given by

$$v_k = u_k + b_k. \tag{6.33}$$

In an equivalent formulation, the Equation 6.32 can be written as

$$
\begin{aligned}
v_k &= \sum_{j=0}^{M} w_{kj}\, x_j, \\
y_k &= \varphi(v_k).
\end{aligned}
\tag{6.34}
$$

In Equation 6.34, a new synapse has been included, with the input $x_0 = +1$ and weight $w_{k0} = b_k$. In this formulation, the effect of the bias term is addressed in the model by adding a new input signal fixed at $+1$ and adding a new synaptic weight equal to the bias b_k. Although the models are different in appearance, they are mathematically equivalent [158].

The activation function, denoted by $\varphi(v)$, defines the output of a neuron in terms of the activation potential v; a few basic types of the activation function includes threshold, piecewise linear, tangent sigmoid, and logarithmic sigmoid.

The manner in which the neurons of an ANN are structured is essentially linked with the learning algorithm used for training the network. Typically, three fundamentally different types of network architectures can be generated [158]; the three types of basic ANN structures are described in the following paragraphs.

Single-layer Feedforward Networks: Typically, the neurons are organized in the form of layers in a layered ANN. The simplest form of a layered network contains an input layer of source nodes which projects data onto an output layer of neurons (i.e., computation nodes). Such a network can be considered to be of the feedforward type with no hidden layer, and is referred to as a single-layer feedforward (SLFF) network. A schematic representation of such a network is shown in Figure 6.2. It should be noted that variations exist in the naming of the layered network based on the total number of layers or number of hidden layers [157, 158].

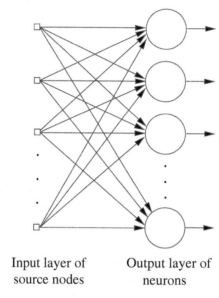

Input layer of   Output layer of
source nodes   neurons

Figure 6.2: Single-layer feedforward neural network.

Multilayer Feedforward Networks: This is a class of the feedforward-type ANN that includes one or more hidden layers consisting of hidden neurons; computation is performed in the hidden layer. The hidden neurons intervene between the external input and the network output in some effective manner. With the addition of one or more hidden layers, the ANN becomes capable of extracting higher order statistics. Typically, the neurons in each layer of the ANN have their input signals from the output signals of the preceding layer only. Studies have indicated that the use of a maximum of two hidden layers is adequate for the approximation of any function. In a binary classification problem, a single neuron in the output layer is adequate; a threshold can be applied to the output value to determine the class.

Recurrent Networks: A recurrent type of ANN is distinguishable from a feedforward type ANN by the presence of at least one feedback loop. In a simple form, a recurrent network may include a single layer of neurons (no hidden layers) with each neuron feeding its output signal back to the inputs of all of the other neurons [158].

In an ANN with a given structure, knowledge representation can be defined by the values taken by the free parameters (i.e., weights and biases) of the network. Typically, a number of labeled examples are presented to the ANN in the form of input-output pairs; the given input-output pairs represent the knowledge about the environment or the task of interest that the given ANN needs to learn through training.

To start with an initial network of a specified structure, all the synaptic weights and thresholds need to be initialized before training. The initialization of the network is critical for classification performance and computational time. Under the assumption that no prior information is available, the synaptic weights and thresholds are typically initialized with values from a uniform distribution whose mean is zero and whose variance is chosen in such a manner that the standard deviation of the activation potential of the neuron lies at the transition between the linear and saturated parts of the sigmoidal activation function [158, 159].

The input samples and the corresponding labeled outputs (targets) are used to train the ANN until the network can closely approximate a function or can perform classification of the given input samples in a specified manner. A typical condition is that the total error between the desired outputs and the actual outputs falls below a predetermined threshold value [159]. Then, samples of unknown categories are presented to the trained ANN for classification.

The training of an ANN classifier is typically performed by the widely used and computationally efficient backpropagation algorithm [158]. The backpropagation algorithm is based upon the error-correction learning rule and can be regarded as a generalization of the adaptive filtering algorithm (i.e., Widrow-Hoff learning rule) for multilayered networks and nonlinear differentiable transfer functions [159]. The backpropagation learning method consists of two passes (i.e., a forward pass and a backward pass) through the layers of the network. In the forward pass, an input pattern is presented to the network, the effect propagates through the layers of the network, and an output response is obtained without affecting the synaptic weights. Then, in the backward pass, the error signal is computed and propagated backward through the network by adjusting the synaptic weights so that the actual response moves closer to the desired response in a statistical sense [158].

In the present work, the simplest representation of a multilayered neural network, an ANN with a single hidden layer included in the structure, was used as one of the classifiers; the hidden layer contains only a single neuron with a tangent-sigmoidal activation function, and it is denoted as SHL-ANN throughout the book. An ANN with a single hidden layer can approximate arbitrarily any function that contains a continuous mapping from one finite space to another [158, 159]. The use of an excessive number of neurons in the hidden layer could result in overfitting (i.e., the ANN has so much information processing capacity that the limited amount of information contained in the training set is not adequate to train all of the neurons in the hidden layers) and significantly increase the time required to train the ANN. Using a significantly smaller number of neurons in the hidden layers could result in underfitting (i.e., there are too few neurons in the hidden layer to detect adequately the signals in a complicated dataset) [158, 159]. Despite the limitation associated (i.e., underfitting), only a single neuron with a tangent-sigmoid activation function was used in the present work for the sake of a simpler structure, for minimal training time, and to achieve classification analogous to logistic classification.

The schematic structure of an SHL-ANN is shown in Figure 6.3. The neurons in the output layer have a pure linear activation function. The Levenberg-Marquardt algorithm [158] with an initial learning rate of 0.05 was selected for fast and robust training with the backpropagation method.

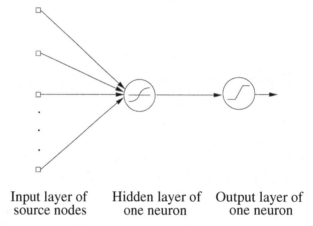

Input layer of   Hidden layer of   Output layer of
source nodes   one neuron   one neuron

Figure 6.3: An SHL (including a single hidden layer) backpropagation ANN with a single neuron in the hidden layer. The hidden neuron has a tangent-sigmoid activation function. The neuron in the output layer has a pure linear activation function.

The Radial Basis Function

The radial basis function (RBF) neural network [158] is a major class of neural network models and can be regarded as a nonlinear mapping between a set of inputs and a set of outputs. In an RBF, the activation of a hidden unit is determined by the distance between the input vector and a prototype vector. Because the mapping functions are nonlinear, it is not necessary to have more than one hidden layer to model any function: a sufficient number of RBF units should be able to approximate the target function [157, 158]. In order to combine the outputs of the hidden RBF units into the network's outputs, a linear combination of the outputs (i.e., a weighted sum of the Gaussian RBFs) could be used to model or approximate any nonlinear function. The standard RBF has an output layer containing dot-product units with identical activation functions [158]. RBF neural networks can be regarded as linear-in-the-parameters models with some unique computational advantages over other architectures of neural networks.

The output, $F : \mathbb{R}^n \to \mathbb{R}$, of the RBF network is

$$F(\mathbf{x}) = \sum_{i=1}^{N} w_i \varphi(||\mathbf{x} - \mathbf{c}_i||), \tag{6.35}$$

where N is the number of neurons in the hidden layer, $\mathbf{c}_i$ is the center vector for neuron i, and w_i are the weights of the linear output neuron. Typically, all inputs are connected to each hidden neuron in an RBF. The norm is usually considered to be the Euclidean distance and the basis function is typically a Gaussian, leading to

$$\varphi\left(||\mathbf{x} - \mathbf{c}_i|| \right) = \exp\left[-\beta\, ||\mathbf{x} - \mathbf{c}_i||^2 \right], \quad \beta > 0. \tag{6.36}$$

The Gaussian basis function is local in the sense that $\lim_{||x|| \to \infty} \varphi(||\mathbf{x} - \mathbf{c}_i||) = 0$; i.e., if the parameters of a neuron are changed, input values that are far away from the center of the neuron will not have a large effect on the output.

RBF networks are considered to be universal approximators on a compact subset of $\mathbb{R}^n$: an RBF network with an adequate number of hidden neurons can approximate any continuous function with arbitrary precision [158]. The weights w_i, the parameter β, and the center vectors $\mathbf{c}_i$ are determined in a manner that optimizes the fit between the output and the input data. Figure 6.4 illustrates the basic structure of an RBF network [158].

The larger the spread of the function φ is, the smoother is the approximation of a function. However, an extremely large spread may lead to many neurons being required to fit a rapidly changing function. On the contrary, too small a spread may cause more neurons to be required to fit a smooth function, and the network might not generalize well [159].

In the present work, all the parameters required for the classifiers used were selected empirically; because of their dependence on the specific inputs, they are not reported here.

6.4.4 SUPPORT VECTOR MACHINES

An SVM is a learning tool based on modern statistical learning theory [160] that gives useful bounds on the generalization capacity of machines for learning tasks. The SVM algorithm constructs a separating hyperplane in the input space; it maps the input space into a higher dimensional feature space through a linear or nonlinear mapping operation [160–163]. SVMs are a set of related supervised learning methods that can be used for classification and regression. Viewing the input data as two sets of vectors in a multidimensional feature space, an SVM constructs a separating hyperplane in that space so as to maximize the margin between the two datasets; the margin is computed by constructing two parallel hyperplanes on each side of the separating hyperplane. Intuitively, a good separation can be achieved by the hyperplane that has the largest distance to the neighboring data points of both classes; typically, the larger the margin, the lower the generalization error of the classifier.

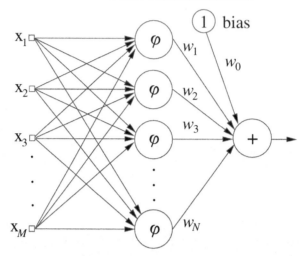

Input layer   Hidden RBF layer   Output layer

Figure 6.4: A radial basis function (RBF) network.

Because of their popularity and effectiveness in pattern classification, a part of the present work included the use of SVMs with RBF kernels. However, due to their highly complex nature and computational time requirement in training, the parts of the results obtained using SVMs are not presented in this book.

6.5   TRAINING AND TEST SETS

Supervised learning is a machine learning technique for creating a function from a set of training samples. The process of using a portion of the entire data to design the classifier is commonly referred to as training of the classifier [139]. The training samples consist of pairs of input objects (typically feature vectors) and desired outputs (targets). The output of the function can be a continuous value (regression), or can predict a class label of the input object (classification). The task of the supervised learner is to predict the value of the function for any valid input object after having seen a number of training samples (i.e., pairs of input feature vectors and the associated output targets) [12, 164].

When a limited number of samples with known classification are available, one needs to find an effective way of using the samples to design or train a classifier, given that the classifier needs to be tested on an independent set of samples with known classification as well [12]. Conventionally, the available observational dataset is often partitioned into two parts: the training data that are used for the identification of a model or classifier including parameter estimation, and the test data that are used for testing the performance of the model or classifier. If the dataset has a small number of samples, proper selection of the training and test sets is of utmost importance.

6.5.1   CROSS-VALIDATION

The division of the training and test data using the 'hold-out' or 'split-sample' method could some-times be subjective, and in some cases, the classifier obtained by the once-partitioned single training dataset could be biased; the classifier structure and the parameters could be highly dependent on how the given dataset was partitioned. The most useful approach to overcome the drawbacks of the hold-out method is to introduce cross-validation [165].

A k-fold cross-validation procedure within the training samples can be employed to estimate the error rate of a classifier. Such a procedure is executed by randomly dividing the training samples into k groups; in each trial, one group is used for testing and the remaining $k - 1$ groups for training, so that every group is used as the test set once. Cross-validation makes good use of the available data as each pattern used is used both as training and test data. Cross-validation is, therefore, especially useful when the amount of available data is insufficient to form the usual training and test partitions required for split-sample training, each of which adequately represents the true distribution of patterns belonging to each class.

6.5.2   THE LEAVE-ONE-OUT METHOD

The case of the most extreme form of cross-validation, where k is set to be equal to the number of training patterns, is known as leave-one-out (LOO) cross-validation, and has been widely studied due to its mathematical simplicity and generalization ability. LOO cross-validation provides an almost unbiased estimate of the generalization ability of a classifier, making it a highly attractive method for the purposes of model selection [166]. The LOO method provides a better estimate of the probability of error because it uses the complete sample population minus one (i.e., $N - 1$) as the training set. Once the classifier is trained, the remaining sample is classified and the procedure is repeated N times until all samples have been classified individually. Since this method uses almost all of the data to classify one sample at a time, its bias is small [165, 167, 168]. However, the main disadvantage of this scheme is that it requires extensive computing time (N number of classifiers have to be designed) [167, 169].

LOO cross-validation is typically recommended for applications where the amount of training data available is strictly limited, such that even a small perturbation of the training data could result in a significant change in the fitted model. In such cases, LOO cross-validation technique would minimize the perturbation to the data in each trial. LOO cross-validation is not preferred for large-scale applications simply due to its extensive computational requirement [165, 166].

However, in some cases, empirical studies have shown that model selection based on k-fold cross-validation could outperform model selection based on LOO cross-validation, because the LOO estimator is known to produce a comparatively high variance. Regardless, for large datasets, the variances of k-fold and LOO estimators are likely to be similar [166].

6.5.3   EFFECTS OF SAMPLE SIZE AND BIAS

The design of classifiers that can accurately distinguish normal and abnormal features is a critical step in the development of CAD algorithms [142]. It has been shown that the performance of a classifier for unknown cases depends on the sample size used for training. When a finite training sample size is used, the performance is pessimistically biased in comparison to that obtained from an infinitely large design sample. In order to design a classifier with a performance generalizable to the large population, a sufficient number of case samples that are representative of the population should be used. However, the availability of case samples is often limited in medical imaging research [142].

In practical applications, the performance of a classifier improves up to a certain point as the number of features is increased and then starts to deteriorate with the addition of further features. For a (linear) multivariate-normal-distribution-based classifier, if the ratio of the sample dimension to the number of features per class is increased above three, the error rate of the classifier approaches the true error rate produced by the classifier based on the minimum probability of error [169, 170], and a number of features of five or more times lower than the number of training samples per class prevents the real performance of the classifier from overestimation [169]. However, as the complexity of the classifier increases, the ratio of the sample dimension to the number of features per class should be increased accordingly; for complex feature spaces and for complex classifiers, a greater ratio of sample size to feature size is required [170].

It has been found that the relative performance of different combinations of classifiers and feature selection methods typically depends on the feature space distributions, the dimensionality, and the available training sample size [171]. The feature selection step could introduce additional biases in the classifier's performance. When the feature selection step is performed using the entire dataset, a positive bias may be introduced in the results obtained by the LOO method or by subsampling and cross-validation methods [170]. To study this effect on the results in the present work, feature selection and pattern classification were performed using several methods to separate the training and testing sets; in particular, the LOO and the 2-fold random subsampling cross-validation methods were applied on the basis of patients, images, and ROIs. In the 2-fold random subsampling cross-validation procedure, approximately 50% of the normal and 50% of the abnormal or TP ROIs, images, or patients were randomly selected for the training procedure (including feature selection) and the rest were used in the testing step, and the procedure was repeated 100 times. In addition, to test the effects of the training set and applicability of the presented methods, two datasets were used: one with the prior images from interval-cancer cases and normal control cases, and the other with prior and detection mammograms of screen-detected cancer cases.

6.6   REMARKS

In this chapter, several methods to perform quantitative analysis of the derived features were presented. The relevant methods of feature selection and pattern classification that are employed in the present work were also described. In the context of CAD, such methods of feature selection, pattern classification, and quantitative evaluation lead toward diagnostic decisions. The performance of the feature selection and pattern classification procedures largely depends on the discriminant capability of the extracted features.

Chapter 7 presents the process of extraction of relevant features for characterization and detection of architectural distortion in prior mammograms. The associated pattern classification experiments conducted and results obtained are described in Chapter 8.

CHAPTER 7

Analysis of Oriented Patterns Related to Architectural Distortion

7.1 FEATURES AND PATTERNS RELATED TO ARCHITECTURAL DISTORTION

The female breast is a complex organ made up of several tissue types, including glandular, fatty, and fibrous tissues, that are positioned over the pectoral muscles of the chest wall [8]. For this reason, mammograms exhibit both random and oriented patterns: random textures arise mainly from the several tissue types and noise, and oriented textures are encountered due to the presence of ducts, vessels, and normal CLS as well as spicules in the presence of spiculated masses or architectural distortion [12]. Although texture patterns could vary significantly between the mammograms of the two breasts of an individual or those of different individuals, any distortion or deviation from the typical oriented patterns could be indicative of pathological conditions. The oriented texture pattern of the normal breast parenchyma generally converges toward the nipple; the presence of architectural distortion distorts the typical pattern, usually with no increased density visible at the center of the distortion [13]. A schematic representation of a node pattern related to architectural distortion is shown in Figure 7.1 (a); two examples of architectural distortion in two mammographic ROIs are also shown in parts (b) and (c) of the same figure.

In the process of detection of architectural distortion, a radiologist performs a visual scan of the mammogram under consideration and searches for specific signs of abnormality, including readily visible signs (i.e., masses and calcifications) and subtle or hard-to-detect signs (i.e., stellate distortion). If a spiculating pattern or distortion of the normal oriented texture pattern appearing as if being pulled toward a central point without the presence of a central density is found, the site is identified as architectural distortion. However, the subtle appearance and the variability in the presentation of architectural distortion create difficulties in the detection process. In this context, several methods are needed to characterize the variety of textural patterns related to architectural distortion in mammograms. As demonstrated in Chapters 4 and 5, the application of Gabor filters and linear phase-portrait analysis leads to the automatic detection of the locations of node-like intersecting or spiculating patterns, including potential sites of architectural distortion; the procedure

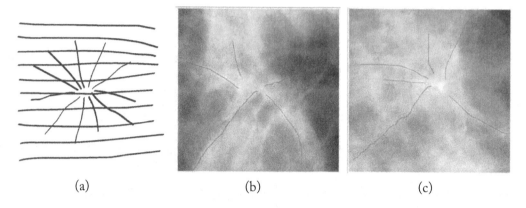

(a)   (b)   (c)

Figure 7.1: (a) A schematic representation of architectural distortion with a node pattern. (b)–(c) Examples of architectural distortion in two mammographic ROIs; the traces (in red) were drawn by a radiologist and correspond to the spiculated components of architectural distortion.

also results in the detection of a number of FP sites. In the present work, the following methods are used for the characterization and detection of architectural distortion:

- Node map analysis using Gabor filters and phase-portrait analysis.

- Power spectral analysis to estimate the FD to characterize the disruption of self-similarity of breast parenchymal patterns due to the presence of architectural distortion, and the analysis of the angular spread of power in the frequency domain for parametric representation of multidirectional oriented patterns.

- Extraction of node, ripple, and wave patterns related to the spiculations and oriented texture of architectural distortion by using Laws' texture energy measures with a geometric transformation.

- Statistical analysis of texture via Haralick's measures.

- Characterization of angular dispersion using the Gabor magnitude response, orientation field angle, coherence, and orientation strength.

- Quantification of angular dispersion using higher order Rényi entropy and Tsallis entropy measures along with Shannon's or Boltzmann-Gibbs's entropy.

For the analysis of the angular spread in the frequency domain and the estimation of FD, methods to transform the 2D Fourier power spectra of the automatically detected ROIs to polar coordinates are presented. For the computation of Laws' texture energy measures, a novel approach is used to transform the rectangular ROIs to polar coordinates for improved discriminant feature extraction and characterization of architectural distortion. Sahiner et al. [172] proposed the rubber band straightening transform (RBST) for texture analysis and characterization of the margins of masses on mammograms. In a similar manner, it is shown that the geometric transformation to polar coordinates leads to improved quantification of the characteristics of intersecting or spiculated parenchymal patterns. The flowchart shown in Figure 7.2 summarizes the feature generation and extraction procedure; the methods are described in detail in the subsequent sections.

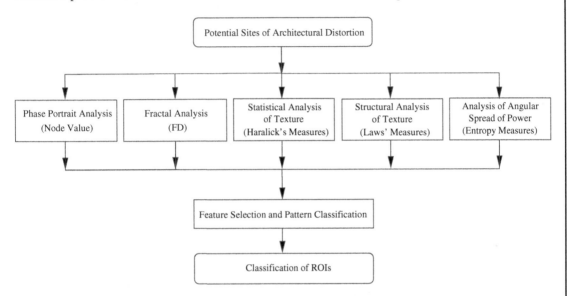

Figure 7.2: Summary of the procedure for the detection of architectural distortion in prior mammograms of interval-cancer cases.

7.2   NODE VALUE

From the results of phase-portrait analysis, as presented in Chapters 4 and 5, a node map was obtained for each mammogram. The peaks in the node map are expected to indicate potential sites of architectural distortion. Hence, the node map was analyzed by rank-ordering the peaks to detect nodes related to the sites of architectural distortion. As discussed in Section 5.2, due to the presence of other structures, intersecting ducts and vessels, and superposition of several texture patterns, the procedure also results in the detection of a number of FP sites. The classification performance of the node values was evaluated using ROC analysis: the A_z obtained was 0.61 for the dataset of interval-cancer cases. The derivation of additional features related to the textural patterns and characteristics of the regions of architectural distortion to assist in separating them from the FPs is described in the following sections.

A square ROI of size 128×128 pixels at 200 μm/pixel (except at the edges of the images) was obtained at each peak in the node map; the center of the node peak was taken as the center of the corresponding ROI. The ROIs were labeled at the locations indicated by the peaks in the node maps, in decreasing order of the node value of the peak, up to a maximum of 30 ROIs per mammogram; see Figure 5.3. The ROIs obtained as above were used for further analysis.

7.3   FREQUENCY-DOMAIN METHODS

The Fourier spectrum of an image with random and oriented texture contains spectral characteristics that may be used for the extraction of important information regarding the objects and patterns present in the image [12].

Let $F(f, \theta)$ be the polar-coordinate representation of the Fourier spectrum of the given image, where $f = \sqrt{u^2 + v^2}$ and $\theta = \tan^{-1}(v/u)$, in terms of the Cartesian frequency coordinates (u, v). The projected function in f or θ may be derived by integrating $F(f, \theta)$ in the other coordinate as

$$F(f) = \int_{\theta=0}^{\pi} F(f, \theta) \, d\theta \qquad (7.1)$$

and

$$F(\theta) = \int_{f=0}^{f_{max}} F(f, \theta) df. \qquad (7.2)$$

The averaging effect due to integration leads to improved visualization of the spectral characteristics of periodic or oriented texture; quantitative features may be derived from $F(f)$ and/or $F(\theta)$ for pattern classification purposes [12]. In this context, frequency-domain methods are used in the present work for characterization of architectural distortion through the estimation of FD and angular spread of power; the methods are described in the following sections.

7.3.1   DESIGN OF GEOMETRIC TRANSFORMATIONS FOR SPECTRAL ANALYSIS

In order to perform analysis in the frequency domain, the 2D Fourier power spectrum of the ROI being processed was obtained with the application of a radial von Hann (also known as Hanning) window [55] and zero padding to the size of 256×256 pixels. The 2D power spectrum $S(u, v)$ in the Cartesian coordinates (u, v) was mapped to the polar coordinates (f, θ) to obtain $S(f, \theta)$, by resampling and computing a weighted average of the four neighbors of each pixel for radial distance f ranging from zero to one-half the sampling frequency and over the range of angle $\theta = [0, 179°]$. Although the power spectrum of ROIs are transformed from rectangular to polar coordinates, the results are presented and processed further in rectangular arrays for ease of computation and representation.

The center (DC frequency) of $S(u, v)$ corresponds to pixel (129, 129). Therefore, for a pixel (f_k, θ_k) in $S(f, \theta)$, $u_k = f_k \cos \theta_k + 129$ and $v_k = 129 - f_k \sin \theta_k$, where (u_k, v_k) are the coordinates of a pixel in $S(u, v)$. See the schematic diagram shown in Figure 7.3 (a). The value of $S(f_k, \theta_k)$

was computed by taking the weighted average of the values of the four neighboring pixels. Defining $u_1 = floor(u_k)$, $u_2 = ceil(u_k)$, $v_1 = floor(v_k)$, and $v_2 = ceil(v_k)$,

$$S(f_k, \theta_k) = \frac{1}{(d_1 + d_2 + d_3 + d_4)} \left[d_1 S(u_1, v_1) + d_2 S(u_1, v_2) + d_3 S(u_2, v_1) + d_4 S(u_2, v_2) \right], \quad (7.3)$$

where

$$
\begin{aligned}
d_1 &= \sqrt{2} - ED11, & ED11 &= \sqrt{[u_k - u_1]^2 + [v_k - v_1]^2}, \\
d_2 &= \sqrt{2} - ED12, & ED12 &= \sqrt{[u_k - u_1]^2 + [v_k - v_2]^2}, \\
d_3 &= \sqrt{2} - ED21, & ED21 &= \sqrt{[u_k - u_2]^2 + [v_k - v_1]^2}, \\
d_4 &= \sqrt{2} - ED22, & ED22 &= \sqrt{[u_k - u_2]^2 + [v_k - v_2]^2}.
\end{aligned}
$$

$$(7.4)$$

See the schematic diagram shown in Figure 7.3 (b). The weights are important to maintain homogeneity and continuity for continuous-to-discrete transformation (i.e., to reduce the estimation error) and also to account for border pixels. Although the Hankel transform [173] and the Fourier transform in polar coordinates have been used for texture analysis and pattern recognition [174], the method described in this book to characterize the angular spread of power is novel and is shown to be effective for the analysis of oriented texture.

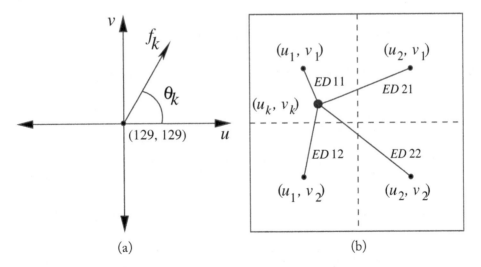

Figure 7.3: Schematic representations of (a) coordinate transformation, and (b) computation of the weighted average of the transformed pixel. The size of the $S(u, v)$ array is 256×256.

7.3.2   FRACTAL ANALYSIS

The fractal property of self-similarity has been observed in many biomedical systems and images [12, 175]. Several studies have suggested that fractal analysis can be used to describe many complex natural phenomena by providing insight into the intrinsic characteristics of the system under investigation [175]. Consequently, fractal analysis has found various applications in medical signal and image processing and analysis [12, 176–178].

For many natural objects or images, the nested recurrence (i.e., self-similarity) of the same structure can appear at smaller and smaller scales for a particular range of scales [176]; the structural or intrinsic characteristics of such objects or images can be quantified by using a measure of fractal geometry, such as FD. FD is related to the complexity in the dimensionality of the object; FD is a nonintegral dimension that quantifies how the given object or pattern fills space and the rate of appearance of additional structural details as the scale of measurement changes. Because FD can be estimated and interpreted in many different ways, it is the most widely used measure in the analysis of medical images among many other fractal measures [177].

Based on the definition of the scale parameters and characteristics of the image being analyzed, different techniques can be formulated for the estimation of FD in order to capture various textural properties [179]. Although several measures have been proposed to estimate the FD of an image, the most frequently used measures of FD are based on the characterization of the self-similarity property or the fractional Brownian motion (fBm) model [177, 178]. For the estimation of the FD of digital images, the most common technique involves the stages of measuring and expressing an image characteristic as a function of the scale parameter for which it is measured, plotting the function in the log-log domain, and computing the slope of the fitted straight line to the generated curve using linear regression. The function in the log-log domain is expected to be linear and the slope of the fitted line is linearly related to FD [38, 75, 76, 175, 178].

In the analysis of fractals and chaos, if the power spectral density (PSD) of a noise process varies as an inverse power of frequency, the process is commonly referred to as $1/f^{\beta}$ noise, where f is the frequency and β is a nonnegative constant scaling factor. $1/f^{\beta}$ noise is observed to occur often in processes found in nature and is a general representation of fBm [180]. In a study conducted by Schepers et al. [180], four methods (relative dispersion, correlation, rescaled range, and power spectral analysis) to estimate the FD of self-affine Brownian noise signals characterized by a specific power spectrum of the form $S(f) \propto f^{-\beta}$ were investigated. The study concluded that the widely known $1/f$ noise is commonly observed in nature; however, the physical reason for this is not well understood. A general model that accounts for both $1/f$ noise and fractals, known as "self-organized criticality," was proposed by Bak et al. [181] for systems with a high number of degrees of freedom.

Many complex theories have been proposed to characterize $1/f^{\beta}$ noise. Lowen and Teich [182] conducted a study on $1/f^{\beta}$ noise and constructed two renewal processes with their PSDs varying as $1/f^{\beta}$. De Los Rios and Zhang [183] showed that the superposition of power spectra with characteristic frequencies suitably distributed in space could be the origin of $1/f^{\beta}$ noise.

In another study, measures of a changing visual environment and perceptual measures of how we see it have shown to exhibit fractal-like multiscale characteristics [184]. In addition, dynamic images of natural scenes and human temporal contrast sensitivity were found to display $1/f^{\beta}$ spectral behavior to some extent [184]. Stŏsić and Stŏsić [185] demonstrated that the human retina represents geometrical multifractals and can be characterized by a hierarchy of exponents rather than a single FD.

In mammography, a few studies have shown that normal and abnormal breast parenchymal patterns exhibit varying degrees of fractal behavior [38, 55, 70, 177, 178, 186]. Because the power spectrum estimation method was observed to provide the most accurate and robust estimates of FD [75, 76, 178], it is widely used in medical image processing and analysis. A frequency domain analysis method, commonly referred to as fractal analysis by circular average (FACA), was proposed by Aguilar et al. [75]; the method eliminates artifacts in the Fourier transform arising from the lack of periodic continuity in real surfaces and profiles, and was shown to provide accurate estimates of the FD of surfaces and profiles. In the present work, the FACA method proposed by Aguilar et al. [75] was used to estimate the FD of the automatically detected ROIs.

Guo et al. [70] used five different methods to estimate the FD, and an SVM to differentiate masses and architectural distortion from normal parenchyma; using FD and lacunarity, the best result obtained for architectural distortion in terms of AUC was 0.875 ± 0.055. Tourassi et al. [55] used FD, estimated using the FACA method, to distinguish between normal tissue patterns and architectural distortion in mammographic ROIs, and obtained an AUC of 0.89. Tourassi et al. concluded that the self-similarity properties of breast parenchymal patterns are disrupted with the presence of architectural distortion, resulting in lower average FD than the average FD of normal breast parenchyma.

7.3.3   ESTIMATION OF FD

In order to estimate the FD, the 2D Fourier power spectrum of the ROI being processed was obtained, including the application of the von Hann window and zero padding to the size of 256 $\times$ 256 pixels. The 2D power spectrum $S(u, v)$ in the Cartesian coordinates (u, v) was mapped to the polar coordinates (f, θ) to obtain $S(f, \theta)$, by using the procedure described in Section 7.3.1 for radial distance f ranging from zero to half the sampling frequency and over the range of angle $\theta = [0, 179°]$. Then, the 2D spectrum $S(f, \theta)$ was transformed into a 1D function $S(f)$, by integrating as a function of the radial distance or frequency f from the zero-frequency point over the range of $\theta = [0, 179°]$ in angle. The geometric transformation as above leads to improved representation and visualization of the spectral characteristics of periodic or spiculated texture [12].

The 1D power spectrum $S(f)$ represents the average value (over all angles) in the 2D power spectrum for a given radial distance (frequency) from the origin (DC), and is considered to be related to the radial frequency f according to the model $S(f) \propto (1/f)^{\beta}$ [181]. Linear regression was applied to a limited frequency range of the 1D spectrum $S(f)$ as a function of the frequency on a log-log scale, excluding points in selected low-frequency and high-frequency regions, to estimate the slope β of the fitted line. The estimated slope is related to FD as [55, 75, 76],

$$FD = \frac{8 - \beta}{2}. \tag{7.5}$$

Selected low- and high-frequency regions were excluded so as to remove the effects of the low-frequency components related to the overall appearance of the image and the large structures present in the image, as well as to prevent the effects of high-frequency noise. The band of frequencies to be excluded (i.e., the nonlinear portion) was selected based on experimentation using synthesized images with known FD, and also using a number of ROIs obtained in the present study. In the present work, the range of f used to fit the linear model is [6, 96] pixels or [0.117, 1.875] mm^{-1}, where the range of [1, 128] pixels corresponds to discrete representation of the frequency range [0, 2.5] mm^{-1}.

Figure 7.4 illustrates three synthetic images with known FD; the estimated FD values using the method described above are given in the caption. Figures 7.5 and 7.6 illustrate the various steps for fractal analysis for a TP ROI and an FP ROI, respectively.

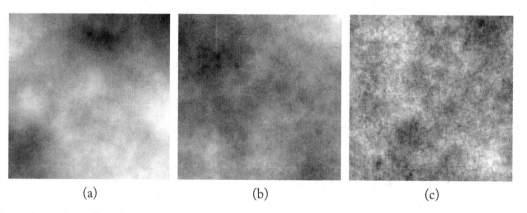

(a)   (b)   (c)

Figure 7.4: Examples of three synthetic images with known FD. Each image is of size 128×128 pixels. (a) True FD= 2.20, estimated FD = 2.29. (b) True FD = 2.50, estimated FD = 2.55. (c) True FD = 2.80, estimated FD = 2.80.

7.3.4 ANGULAR SPREAD OF POWER

Although several studies have been reported on fractal analysis using the Fourier power spectrum, the angular spread of power is often overlooked; only the Fourier spectral magnitude as a function of radial frequency averaged over all angles (as in the FACA method) is used, and the information related to the angular spread of power in the frequency domain is usually ignored. In the context of the present work, the angular spread of power in the Fourier spectrum is of interest and could be used to generate features for the detection of spiculated patterns related to architectural distortion.

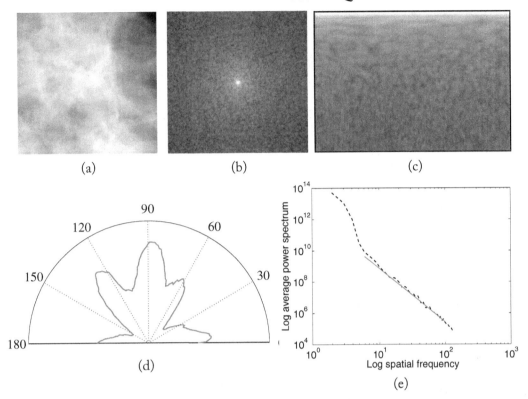

Figure 7.5: (a) A 128×128-pixel TP ROI with architectural distortion; pixel size $= 200$ μm. (b) The 2D Fourier log-power spectrum $S(u, v)$ obtained after applying the von Hann window and zero padding the ROI to 256×256 pixels. (c) The power spectrum in the (f, θ) space. The horizontal axis corresponds to angle θ from $0°$ to $179°$ and the vertical axis corresponds to radial frequency from 0.02 mm^{-1} to 2.5 mm^{-1}. The top-left corner pixel corresponds to frequency of 0.02 mm^{-1} and angle of $0°$. A black frame has been applied to the spectrum. (d) Angular spread of power, $S(\theta)$. Entropy $H = 7.36$; variance σ_θ^2 $= 0.56E^{-5}$; weighted standard deviation $SD_\theta = 0.81$. (e) The 1D power spectrum $S(f)$ plotted on a log-log scale as a function of radial frequency f. The linear fit is also shown, which resulted in the FD of 2.28 for the TP ROI. Reproduced with permission from: S. Banik, R. M. Rangayyan, and J. E. L. Desautels. "Detection of Architectural Distortion in Prior Mammograms." *IEEE Transactions on Medical Imaging*, vol. 30(2), pp. 279–294, February 2011. © IEEE.

To characterize the angular spread of power in the frequency domain, the geometrically transformed 2D Fourier power spectrum $S(f, \theta)$ was transformed into a 1D function $S(\theta)$, by integrating as a function of the angle θ (for the range $[0, 179°]$) from the zero-frequency point over radial distance $f = [1, 128]$ pixels. Selected low- and high-frequency regions were excluded in the same

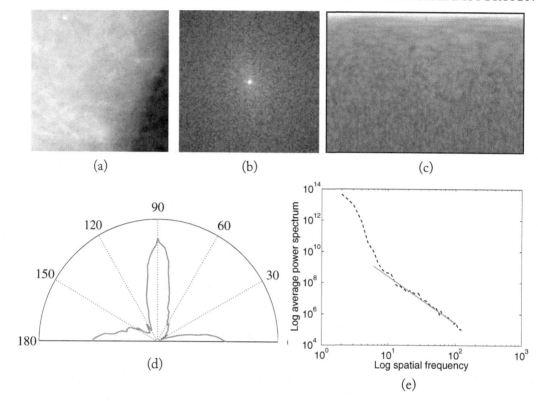

Figure 7.6: (a) A 128×128-pixel FP ROI; pixel size $= 200$ μm. The ROI caused an FP node due to intersecting normal structures. (b) The 2D Fourier log-power spectrum $S(u, v)$ obtained after applying the von Hann window and zero padding the ROI to 256×256 pixels. (c) The power spectrum in the (f, θ) space. The horizontal axis corresponds to angle θ from $0°$ to $179°$ and the vertical axis corresponds to radial frequency from 0.02 mm^{-1} to 2.50 mm^{-1}. The top-left corner pixel corresponds to frequency of 0.02 mm^{-1} and angle of $0°$. A black frame has been applied to the spectrum. (d) Angular spread of power, $S(\theta)$. Entropy $H = 7.10$; variance $\sigma_\theta^2 = 0.16E^{-4}$; weighted standard deviation $SD_\theta = 0.89$. (e) The 1D power spectrum $S(f)$ plotted on a log-log scale as a function of radial frequency f. The linear fit is also shown, which resulted in the FD of 2.52 for the FP ROI. Reproduced with permission from: S. Banik, R. M. Rangayyan, and J. E. L. Desautels. "Detection of Architectural Distortion in Prior Mammograms." *IEEE Transactions on Medical Imaging*, vol. 30(2), pp. 279–294, February 2011. © IEEE.

manner as described above. Due to the presence of spiculations radiating at several angles, TP ROIs are expected to have a large angular spread of power. On the other hand, most FP ROIs were observed to contain a few intersecting ligaments, ducts, and vessels with the related spectral power limited to a small number of angular bands.

To quantify the angular spread of power, three features were computed from $S(\theta)$ after being normalized to have unit sum: Shannon's or Boltzmann-Gibbs's entropy as

$$H = - \sum_{\theta} S(\theta) \log_2 S(\theta), \tag{7.6}$$

variance of $S(\theta)$ as

$$\sigma_{\theta}^2 = \frac{1}{N-1} \sum_{\theta} [S(\theta) - \overline{S(\theta)}]^2, \tag{7.7}$$

and the standard deviation of θ weighted by $S(\theta)$ as

$$SD_{\theta} = \sqrt{\sum_{\theta} [\theta - \bar{\theta}]^2 S(\theta)}. \tag{7.8}$$

Here,

$$\bar{\theta} = \sum_{\theta} \theta S(\theta), \tag{7.9}$$

$N = 180$ is the number of bins used to calculate $S(\theta)$, and

$$\overline{S(\theta)} = \frac{1}{N} \sum_{\theta} S(\theta) = \frac{1}{N}. \tag{7.10}$$

However, the variance σ_{θ}^2 calculated as above is highly correlated with the entropy H (correlation coefficient = 0.995), with both resulting in the same A_z value (0.64). The variance σ_{θ}^2 was not adding any extra discriminative information; as such, it was not selected by the feature selection procedure used in an earlier study [87]. In addition, the weighted standard deviation SD_{θ} was observed to have poor discriminant power ($A_z = 0.51$) with a high p-value (0.81), and was not selected by the feature selection procedure as well. Consequently, these two features were not used in the present work; instead, two higher order entropy measures, Tsallis and Rényi entropy, were used for quantification of the angular spread in the frequency domain. The Tsallis and Rényi entropy measures are discussed in Sections 7.6.1 and 7.6.2, respectively.

Figures 7.5 and 7.6 illustrate the various steps for fractal analysis and the estimation of the angular spread of power in the frequency domain for a TP ROI and an FP ROI, respectively. Figure 7.5 (d) indicates the existence of multidirectional spiculating patterns for the TP ROI, whereas in Figure 7.6 (d), the spread of power is limited to a small number of angular bands for the FP ROI.

7.4   ANALYSIS OF TEXTURE

In image processing and analysis, texture can be considered as a function of the spatial variation in pixel intensities or gray-level values and is one of the important characteristics of images [12].

Texture analysis is an important and useful area of study in machine vision: recognition of objects or classification of normal and abnormal regions can be performed using the textural properties of a given image. Although various types of texture patterns can be observed in medical images, oriented texture is commonly encountered due to the extensive presence of muscles, ligaments, blood vessels, ducts, nerves, and other piecewise linear structures. Analysis of the dominant orientation of texture patterns often provides important information regarding the object under consideration, and could be used in characterizing or deriving discriminant features for the object. On the other hand, analysis of random patterns associated with certain regions in an image could also help in generating useful features for the detection abnormalities. As a result, multiple methods are required to characterize the variety of texture found in mammograms. In the following sections, various methods for texture analysis used in the present work are described.

7.4.1 STATISTICAL ANALYSIS USING HARALICK'S MEASURES

For statistical analysis of texture, in particular random texture, Haralick's texture features [91, 92] are the most commonly used measures in image analysis and machine vision. Haralick et al. proposed 14 statistical measures for performing analysis of random texture; the measures are computed based upon the moments of a joint PDF estimated using the joint occurrence or cooccurrence of gray levels, commonly referred to as the gray-level cooccurrence matrix (GCM). GCMs are also known as spatial gray-level dependence (SGLD) matrices [12].

A GCM of an image (or ROI) is formed by considering the corresponding distribution of cooccurring values at a given spatial offset. The GCM $P_{(d,\theta)}(l_1, l_2)$ corresponds to the probability of occurrence of the pair of gray levels (l_1, l_2) separated by the given distance d at the angle θ in the image. To form GCMs, the gray-level co-occurrence counts or probabilities based on the spatial relations of pixels at different distances (specified by d) and angular directions (specified by θ) are mapped by scanning the image from left to right and top to bottom.

In natural images, objects are typically comprised of different gray-level values that do not vary extensively within small distances in a particular object. Consequently, neighboring pixels in natural images possess nearly the same or correlated gray-level values, causing the corresponding GCMs to have large values along and around the main diagonal (for $l_1 \approx l_2$), and low values away from the diagonal. Because it may not be possible to pair the pixels in a few rows or columns at the borders of an image with another pixel according to the chosen parameters (d, θ), the number of pairs of pixels that can be formed will always be less than MN for an image of size $M \times N$.

GCMs are commonly constructed for unit-pixel distances ($d = 1$) and the four angles of $\theta = 0°, 45°, 90°$, and $135°$ in typical texture analysis applications. The average of the four GCMs is computed when directional analysis of features at certain angles is not required. The 14 statistical measures of texture, as proposed by Haralick et al., are based upon normalized GCMs that can be written as:

$$p(l_1, l_2) = \frac{P(l_1, l_2)}{\sum_{l_1=0}^{L-1} \sum_{l_2=0}^{L-1} P(l_1, l_2)}, \tag{7.11}$$

where L is the number of gray levels in the image and the subscript (d, θ) has been dropped to reduce complexity in representation.

Digital or digitized mammograms are typically acquired at the resolution of about $50\mu m$ per pixel, with 4,096 gray levels represented using 12 bpp. However, the computational effort required to construct the GCM of a 12-bpp image would be excessively high for practical applications; furthermore, most pairs of gray levels would occur with low or negligible numbers of incidence, resulting in difficulties in the derivation of reliable statistics. Therefore, in order to avoid sparse GCMs, it is advantageous to reduce the image to 256 gray levels (i.e., 8 bpp). In the present work, the ROIs obtained were quantized at 8 bpp before computing Haralick's texture measures.

All of the 14 features proposed by Haralick et al. are used in the present study and are listed in Table 7.1. These statistical measures of texture reveal properties about the spatial distribution of gray levels in the image. Further details of the characteristics of the 14 statistical measures of texture based upon the GCM can be found in the publications of Haralick et al. [91, 92]. Some of the texture features defined in Table 7.1 have values much greater than unity, whereas some of the features have values far less than unity. Normalization to a predefined range, such as [0, 1], over the dataset to be analyzed, could be beneficial in further analysis [12].

Texture analysis, using some or all of Haralick's 14 texture features, is a popular approach for the analysis and classification of many medical images, including breast masses and tumors seen in mammograms. Haralick's measures facilitate statistical analysis of texture and could help in the process of characterization of architectural distortion. Therefore, in the present work, Haralick's texture measures have been chosen for the analysis of the texture related to architectural distortion.

Table 7.1: List of Haralick's texture features used to characterize texture in mammograms.

Feature No.	Feature Name	Definition
1	Energy	$HT_1 = \sum_{l_1=0}^{L-1} \sum_{l_2=0}^{L-1} p^2(l_1, l_2).$
2	Contrast	$HT_2 = \sum_{k=0}^{L-1} k^2 \underbrace{\sum_{l_1=0}^{L-1} \sum_{l_2=0}^{L-1}}_{\|l_1 - l_2\|=k} p(l_1, l_2).$

Table 7.1: *Continued.*

Feature No.	Feature Name	Definition
3	Correlation	$HT_3 = \frac{1}{\sigma_x\,\sigma_y}\left[\sum_{l_1=0}^{L-1}\sum_{l_2=0}^{L-1} l_1\,l_2\,p(l_1,l_2) - \mu_x\,\mu_y\right]$ where μ_x and μ_y are the means and σ_x and σ_y are the standard deviation values of p_x and p_y, respectively. The marginal probabilities are: $$p_x(l_1) = \sum_{l_2=0}^{L-1} p(l_1,l_2)$$ and $$p_y(l_2) = \sum_{l_1=0}^{L-1} p(l_1,l_2).$$
4	Sum of Squares	$HT_4 = \sum_{l_1=0}^{L-1}\sum_{l_2=0}^{L-1} (l_1 - \mu_f)^2\,p(l_1,l_2)$, where μ_f is the mean gray level of the image.
5	Inverse Difference Moment	$HT_5 = \sum_{l_1=0}^{L-1}\sum_{l_2=0}^{L-1} \frac{1}{1+(l_1-l_2)^2}\,p(l_1,l_2).$
6	Sum Average	$HT_6 = \sum_{k=0}^{2(L-1)} k\,p_{x+y}(k)$ where p_{x+y} is given by $$p_{x+y}(k) = \underbrace{\sum_{l_1=0}^{L-1}\sum_{l_2=0}^{L-1}}_{l_1+l_2=k} p(l_1,l_2).$$
7	Sum Variance	$HT_7 = \sum_{k=0}^{2(L-1)} (k - HT_6)^2\,p_{x+y}(k).$
8	Sum Entropy	$HT_8 = -\sum_{k=0}^{2(L-1)} p_{x+y}(k)\,\log_2\left[p_{x+y}(k)\right].$
9	Entropy	$HT_9 = -\sum_{l_1=0}^{L-1}\sum_{l_2=0}^{L-1} p(l_1,l_2)\,\log_2[p(l_1,l_2)].$

Table 7.1: *Continued.*

Feature No.	Feature Name	Definition
10	Difference Variance	$HT_{10} = \sum_{k=0}^{2(L-1)} \left(k - \sum_{k=0}^{2(L-1)} k\, p_{x-y}(k) \right)^2 p_{x-y}(k)$ where p_{x-y} is given by $$p_{x-y}(k) = \underbrace{\sum_{l_1=0}^{L-1} \sum_{l_2=0}^{L-1} p(l_1, l_2)}_{|l_1 - l_2| = k}.$$
11	Difference Entropy	$HT_{11} = - \sum_{k=0}^{L-1} p_{x-y}(k) \, \log_2 \left[p_{x-y}(k) \right].$
12	Information-theoretic Measures of Correlation[1]	$$HT_{12} = \frac{H_{xy} - H_{xy1}}{\max[H_x, H_y]}$$ where $H_{xy} = HT_9$; H_x and H_y are the entropies of p_x and p_y, respectively; $$H_{xy1} = - \sum_{l_1=0}^{L-1} \sum_{l_2=0}^{L-1} p(l_1, l_2) \, \log_2 \left[p_x(l_1)\, p_y(l_2) \right],$$ and $$H_{xy2} = - \sum_{l_1=0}^{L-1} \sum_{l_2=0}^{L-1} p_x(l_1)\, p_y(l_2) \, \log_2 \left[p_x(l_1)\, p_y(l_2) \right].$$
13	Information-theoretic Measures of Correlation[2]	$HT_{13} = \left\{ 1 - \exp[-2\,(H_{xy2} - H_{xy})] \right\}^{\frac{1}{2}}.$
14	Maximal Correlation Coefficient	$HT_{14} = $ *(Second largest Eigenvalue of Q)*$^{1/2}$ where Q is computed as $$Q(l_1, l_2) = \sum_{k=0}^{L-1} \frac{p(l_1, k)\, p(l_2, k)}{p_x(l_1)\, p_y(k)}.$$

7.4.2 STRUCTURAL ANALYSIS OF TEXTURE USING LAWS' ENERGY MEASURES

Laws' texture energy measures are based on convolution kernels that emphasize specific structural patterns, and could be used to generate useful features related to the intersecting tissue structures, spiculations, and node-like patterns of architectural distortion. Laws [90] defined several 1D and 2D convolution kernels to classify each pixel in an image based on measures of local "texture energy." The texture energy features represent the amounts of variation within a sliding window applied to several filtered versions of the given image [12].

The basic operators in Laws' method are:

$$L3 = [1,\ 2,\ 1],$$

$$E3 = [-1,\ 0,\ 1],$$

and

$$S3 = [-1,\ 2,\ -1],$$

where $L3$, $E3$, and $S3$ perform center-weighted averaging, symmetric first differencing (edge detection), and second differencing (spot detection), respectively. Nine 3×3 masks can be generated by multiplying the transposes of the three operators with their direct versions. For example, the result of $L3^T E3$ gives one of the 3×3 Sobel masks as

$$\begin{bmatrix} -1 & 0 & 1 \\ -2 & 0 & 2 \\ -1 & 0 & 1 \end{bmatrix}.$$

Nine 1D operators of length five pixels can be generated by convolving $L3$, $E3$, and $S3$ operators in different combinations; for example

$$
\begin{aligned}
L5 &= L3 * L3 &= [1,\ 4,\ 6,\ 4,\ 1], \\
R5 &= S3 * S3 &= [1,\ -4,\ 6,\ -4,\ 1], \\
W5 &= -E3 * S3 &= [-1,\ 2,\ 0,\ -2,\ 1], \\
E5 &= L3 * E3 &= [-1,\ -2,\ 0,\ 2,\ 1], \\
S5 &= -E3 * E3 &= [-1,\ 0,\ 2,\ 0,\ -1],
\end{aligned}
$$

where $*$ represents 1D convolution. The operators $L5$, $R5$, $W5$, $E5$, and $S5$ can be used to perform the detection of features related to local average, ripples, waves, edges, and spots.

For the analysis of 2D images, the 1D convolution operators given above may be used to generate 2D convolution masks of size 5×5 pixels to emphasize center-weighted local average $(L5L5 = L5^T L5)$, ripples $(R5R5 = R5^T R5)$, and wave patterns $(W5W5 = W5^T W5)$, respectively. The $L5L5$, $W5W5$, and $R5R5$ operators in 2D are shown in Figure 7.7.

Because architectural distortion is expected to include spiculations radiating from a point at or near the center of the corresponding ROI and node-like patterns, in the present work it is

$$\begin{bmatrix} 1 & 4 & 6 & 4 & 1 \\ 4 & 16 & 24 & 16 & 4 \\ 6 & 24 & 36 & 24 & 6 \\ 4 & 16 & 24 & 16 & 4 \\ 1 & 4 & 6 & 4 & 1 \end{bmatrix} \qquad \begin{bmatrix} 1 & -2 & 0 & 2 & -1 \\ -2 & 4 & 0 & -4 & 2 \\ 0 & 0 & 0 & 0 & 0 \\ 2 & -4 & 0 & 4 & -2 \\ -1 & 2 & 0 & -2 & 1 \end{bmatrix}$$

$$L5L5 \qquad\qquad\qquad W5W5$$

$$\begin{bmatrix} 1 & -4 & 6 & -4 & 1 \\ -4 & 16 & -24 & 16 & -4 \\ 6 & -24 & 36 & -24 & 6 \\ -4 & 16 & -24 & 16 & -4 \\ 1 & -4 & 6 & -4 & 1 \end{bmatrix}$$

$$R5R5$$

Figure 7.7: Laws' operators for 2D convolution.

hypothesized that Laws' 5×5 convolution masks for the wave detector ($W5W5$), the ripple detector ($R5R5$), and the center-weighted local average ($L5L5$) should provide discriminant information. The application of the edge detector ($E5E5$) and the spot detector ($S5S5$) was found to be irrelevant for the patterns of architectural distortion, and was not used for further analysis.

The rectangular ROIs automatically extracted from mammograms and the corresponding Gabor magnitude responses were geometrically transformed to polar coordinates before applying Laws' method; the method of transformation is described in Section 7.4.3. The geometrically transformed ROIs and the transformed Gabor magnitude responses were convolved with Laws' 5×5 convolution masks designed for the detection of waves ($W5W5$), ripples ($R5R5$), and the center-weighted local average ($L5L5$). In addition, two other masks, $W5W5$ rotated by 45° and $R5R5$ rotated by 45°, were used. Following the application of the selected filters, texture energy measures were derived from each of the filtered images by computing the average of the squared values in a 15×15 sliding window. The results of application of the methods to a test pattern, a sample TP ROI, and a sample FP ROI are shown in Figures 7.8, 7.9, and 7.10, respectively.

Finally, the sum of each of the energy measures normalized by the area of the transformed image was used to derive 10 features (five from the transformed ROIs and five from the transformed Gabor magnitude responses) for feature selection and pattern classification; the features are listed in Table 7.2.

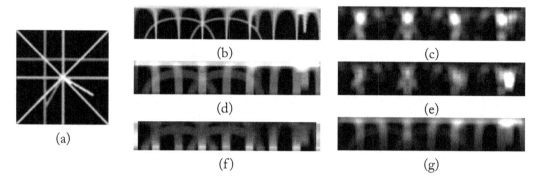

Figure 7.8: (a) A test image of size 128×128 pixels including patterns simulating spiculations and intersecting linear patterns. (b) The ROI transformed to the polar (r, θ) coordinates. (c)–(g) Results of application of Laws' $W5W5$, rotated $W5W5$, $R5R5$, rotated $R5R5$, and $L5L5$ masks, respectively, to the image shown in part (b) and computing the average of the squared values in a 15×15 sliding window. For all of the transformed images, the horizontal axis corresponds to angle $0° \leq \theta \leq 359°$ and the vertical axis corresponds to radial distance $2 \leq r \leq 63$ pixels from the center of the original image. The top-left corner pixel corresponds to $(r, \theta) = (2, 0°)$. The values of the images have been scaled for the purpose of illustration. Reproduced with permission from: S. Banik, R. M. Rangayyan, and J. E. L. Desautels. "Detection of Architectural Distortion in Prior Mammograms." *IEEE Transactions on Medical Imaging*, vol. 30(2), pp. 279–294, February 2011. © IEEE.

7.4.3   DESIGN OF GEOMETRIC TRANSFORMATIONS FOR THE ANALYSIS OF ORIENTED PATTERNS

The ROIs were mapped to the polar (r, θ) space from the Cartesian (x, y) space, for radial distances ranging from two to one-half of the smaller dimension, over the range of angles $[0°, 359°]$, and by taking weighted average of the four neighbors of each pixel. Similar to the method presented in Section 7.3.1, the process is described as follows.

Let $I(x, y)$ be a rectangular ROI of size $M \times N$, which is to be transformed to an image $J(r, \theta)$, where r ranges from 2 to $\lceil \frac{(M, N)}{2} \rceil - 1$ (one-half of the smaller of the two dimensions rounded up to the smallest following integer minus one), and θ ranges from $0°$ to $359°$. Let (x_0, y_0) be the center of $I(x, y)$. Then, for a pixel in the image $J(r, \theta)$ corresponding to radius r_z and angle θ_z, the relationship can be expressed as $x_z = r_z \cos \theta_z + x_0$ and $y_z = -r_z \sin \theta_z + y_0$, where $I(x_z, y_z)$ is the pixel at (x_z, y_z). See the schematic diagram shown in Figure 7.11 (a). The value of $J(r_z, \theta_z)$ was computed by taking the weighted average of the values of the four neighboring corner pixels of $I(x_z, y_z)$. Defining, $x_1 = floor(x_z)$, $x_2 = ceil(x_z)$, $y_1 = floor(y_z)$, and $y_2 = ceil(y_z)$,

$$J(r_k, \theta_k) = \frac{1}{(d_1 + d_2 + d_3 + d_4)} [d_1 I(x_1, y_1) + d_2 I(x_1, y_2) + d_3 I(x_2, y_1) + d_4 I(x_2, y_2)],$$

$$(7.12)$$

Table 7.2: List of features with A_z values for the interval-cancer dataset. FD: fractal dimension; LT1–LT10: Laws' texture energy measures; HT1–HT14: Haralick's texture features; AS1–AS15: Features related to the angular spread of power. The Rényi entropy used is of order 8 and Tsallis entropy is of order 2. Ranking is based on the A_z value. Machine precision: $\epsilon = 2.2204E - 016$. *Continues.*

Category	Feature No.	Feature description	A_z	p-value	Ranking
Node map analysis	Node	Node value	0.61	$1.7638\,E - 009$	22
Fractal analysis	FD	Fractal dimension	0.59	$1.8793E - 004$	24

where

$$d_1 = \sqrt{2} - ED11, \quad ED11 = \sqrt{[x_z - x_1]^2 + [y_z - y_1]^2},$$
$$d_2 = \sqrt{2} - ED12, \quad ED12 = \sqrt{[x_z - x_1]^2 + [y_z - y_2]^2},$$
$$d_3 = \sqrt{2} - ED21, \quad ED21 = \sqrt{[x_z - x_2]^2 + [y_z - y_1]^2},$$
$$d_4 = \sqrt{2} - ED22, \quad ED22 = \sqrt{[x_z - x_2]^2 + [y_z - y_2]^2}.$$

$$(7.13)$$

See the schematic diagram shown in Figure 7.11 (b). The weights are important to maintain homogeneity and continuity for continuous-to-discrete transformation and also to account for border pixels.

The geometric transformation as described above converts spiculated patterns into ripple or wave patterns as illustrated in Figure 7.8. ROC analysis of related features confirmed that better performance is achieved by using the ROIs transformed to the polar coordinates than the original ROIs in the Cartesian coordinates, in discriminating between TP and FP ROIs. It should be noted that, although ROIs are transformed from rectangular to polar coordinates, the results are presented and processed further in rectangular arrays for ease of computation and representation. A limitation of the transformation is that it only considers the pixels within circle of radius of r ranging from 2 to $\lceil \frac{(M,N)}{2} \rceil - 1$ in the ROI for transformation; the pixels outside the circle are not included for transformation. However, the rectangular ROIs are centered at the peaks in the node map and are expected to contain the parts of architectural distortion within the circle described above; the corner pixels outside the circle are not expected to contain any important information related to the oriented or spiculating patterns of architectural distortion.

The geometric transformation of the ROI being processed so as to convert spiculating or radiating patterns to waves or ripples is an important prerequisite to the derivation of Laws' texture

Table 7.2: *Continued.*

Category	Feature No.	Feature description	A_z	p-value	Ranking
Laws' texture energy	LT1	$W5W5$ on transformed Gabor magnitude	0.63	$7.1652E-008$	16
	LT2	Rotated $W5W5$ on transformed Gabor magnitude	0.56	$8.9615E-008$	29
	LT3	$R5R5$ on transformed Gabor magnitude	0.61	$3.3171E-005$	20
	LT4	Rotated $R5R5$ on transformed Gabor magnitude	0.58	$9.0572E-010$	26
	LT5	$L5L5$ on transformed Gabor magnitude	0.56	$7.3351E-006$	30
	LT6	$W5W5$ on transformed ROI	0.57	0.6837	28
	LT7	Rotated $W5W5$ on transformed ROI	0.65	$9.6034E-014$	10
	LT8	$R5R5$ on transformed ROI	0.60	$6.3091E-004$	23
	LT9	Rotated $R5R5$ on transformed ROI	0.61	$7.4237E-005$	21
	LT10	$L5L5$ on transformed ROI	0.65	$1.6320E-014$	9
Haralick's texture	HT1	Energy (HT_1)	0.59	$7.1925E-007$	25
	HT2	Contrast (HT_2)	0.55	$9.7649E-005$	32
	HT3	Correlation (HT_3)	0.53	$1.0157E-005$	35
	HT4	Sum of Squares (HT_4)	0.51	0.0331	37
	HT5	Inverse Difference Moment (HT_5)	0.56	$2.6209E-004$	31
	HT6	Sum Average (HT_6)	0.63	$6.9300E-013$	17
	HT7	Sum Variance (HT_7)	0.51	0.0338	38
	HT8	Sum Entropy (HT_8)	0.53	0.0118	34
	HT9	Entropy (HT_9)	0.54	0.0029	33
	HT10	Difference Variance (HT_{10})	0.57	$4.9585E-005$	27

Table 7.2: *Continued.*

Category	Feature No.	Feature description	A_z	p-value	Ranking
	HT11	Difference Entropy (HT_{11})	0.50	0.7504	39
	HT12	Information-theoretic Measure of Correlation[1] (HT_{12})	0.50	0.4340	40
	HT13	Information-theoretic Measure of Correlation[2] (HT_{13})	0.52	$1.3187E - 004$	36
	HT14	Maximal Correlation Coefficient (HT_{14})	0.50	$7.5380E - 004$	41
Angular spread	AS1	Shannon's Entropy: power in the frequency domain	0.64	$2.2615E - 013$	13
	AS2	Shannon's Entropy: Gabor magnitude response	0.68	$2.2204E - 016$	5
	AS3	Shannon's Entropy: Gabor angle response	0.63	$4.3465E - 011$	18
	AS4	Shannon's Entropy: coherence	0.68	$2.2204E - 016$	6
	AS5	Shannon's Entropy: orientation strength	0.62	$2.8177E - 010$	19
	AS6	Tsallis Entropy: power in the frequency domain	0.64	$1.9937E - 012$	12
	AS7	Tsallis Entropy: Gabor magnitude response	0.68	$1.9393E - 012$	3
	AS8	Tsallis Entropy: Gabor angle response	0.64	$5.3130E - 012$	14
	AS9	Tsallis Entropy: coherence	0.68	$1.6726E - 012$	4
	AS10	Tsallis Entropy: orientation strength	0.63	$1.1146E - 011$	15
	AS11	Rényi Entropy: power in the frequency domain	0.64	$1.6653E - 015$	11
	AS12	Rényi Entropy: Gabor magnitude response	0.69	$2.2204E - 016$	1
	AS13	Rényi Entropy: Gabor angle response	0.67	$2.2204E - 016$	8
	AS14	Rényi Entropy: coherence	0.69	$2.2204E - 016$	2
	AS15	Rényi Entropy: orientation strength	0.67	$2.2204E - 016$	7

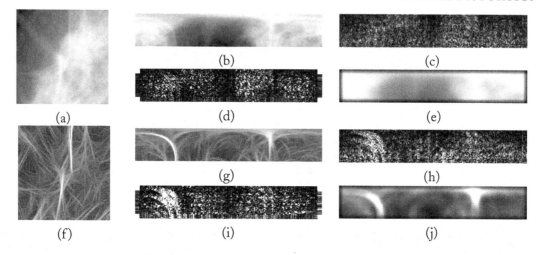

Figure 7.9: (a) A 128 × 128-pixel TP ROI with architectural distortion. (b) The ROI transformed to the polar (r, θ) coordinates. (c)–(e) Results of application of Laws' $W5W5$, $R5R5$, and $L5L5$ masks, respectively, to the image shown in part (b) and computing the average of the squared values in a 15 × 15 sliding window. (f) The Gabor magnitude response for the ROI shown in part (a). (g)–(j) Procedure repeated as above for the Gabor magnitude response shown in part (f). For all of the transformed images, the horizontal axis corresponds to angle $0° \leq \theta \leq 359°$ and the vertical axis corresponds to radial distance $2 \leq r \leq 63$ pixels from the center of the original ROI. The top-left corner pixel corresponds to $(r, \theta) = (2, 0°)$. The values of the images have been scaled for the purpose of illustration. Reproduced with permission from: S. Banik, R. M. Rangayyan, and J. E. L. Desautels. "Detection of Architectural Distortion in Prior Mammograms." *IEEE Transactions on Medical Imaging*, vol. 30(2), pp. 279–294, February 2011. © IEEE.

energy measures for the characterization of architectural distortion, and represents an important and novel contribution of the present work.

7.5   CHARACTERIZATION OF ANGULAR DISPERSION

In a preliminary study related to the present work, characterization of the angular spread of power in the frequency domain was shown to be effective in the detection of architectural distortion [84, 187, 188]. In this context, it is hypothesized that analysis of the angular histograms of the Gabor magnitude response, orientation angle, coherence, and orientation strength could reveal important information regarding the presence of architectural distortion. It should be noted that only the Gabor orientation field (angle response) was used in phase-portrait analysis; the Gabor magnitude response could also aid in the detection of architectural distortion as demonstrated in the following sections.

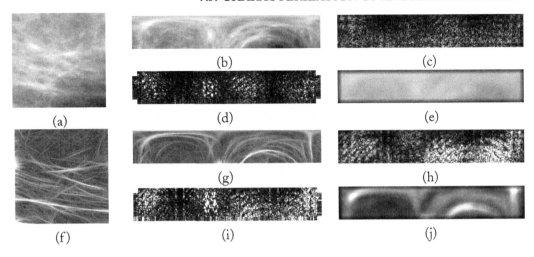

Figure 7.10: (a) A 128 × 128-pixel FP ROI. (b) The ROI transformed to the polar (r, θ) coordinates. (c)–(e) Results of application of Laws' $W5W5$, $R5R5$, and $L5L5$ masks, respectively, to the image shown in part (b) and computing the average of the squared values in a 15 × 15 sliding window. (f) The Gabor magnitude response for the ROI shown in part (a). (g)–(j) Procedure repeated as above for the Gabor magnitude response shown in part (f). For all of the transformed images, the horizontal axis corresponds to angle $0° \leq \theta \leq 359°$ and the vertical axis corresponds to radial distance $2 \leq r \leq 63$ pixels from the center of the original ROI. The top-left corner pixel corresponds to $(r, \theta) = (2, 0°)$. The values of the images have been scaled for the purpose of illustration. Reproduced with permission from: S Banik, R M Rangayyan, and J E L Desautels. "Digital Image Processing and Machine Learning Techniques for the Detection of Architectural Distortion in Prior Mammograms." In K. Suzuki, Ed., *Machine Learning in Computer-aided Diagnosis: Medical Imaging Intelligence and Analysis*, pp. 24–63, IGI Global, Hershey, PA, January 2012. © IGI Global.

The output of the bank of 180 real Gabor filters (as described in Section 3.2.1) was used to derive the magnitude response, orientation field, coherence, and orientation strength images. The magnitude response and angle of the Gabor filter with the highest output for a given pixel were used to construct the magnitude response and orientation field images. The derivation of coherence and orientation strength is described in the following sections.

7.5.1   COHERENCE

In the method for computing the coherence, the orientation information at each pixel is represented by a single pair of magnitude and orientation values. The method is based on each pixel's neighborhood, giving the dominant orientation in an average sense and the degree of alignment

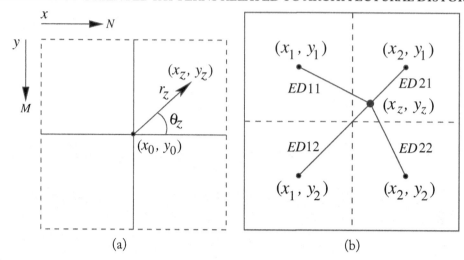

(a)   (b)

Figure 7.11: Schematic representations of (a) coordinate transformation, and (b) computation of the weighted average of the transformed pixel.

of the orientation information for each pixel in the neighborhood with respect to the dominant orientation [88, 189]. The dominant orientation is computed as the orientation that maximizes the coherence and is described as follows.

Let $G(m, n)$ and $\theta(m, n)$ represent the gradient magnitude and gradient orientation at the point (m, n) in an image, respectively, and $P \times P$ be the size of the neighborhood around (p, q) used for computing the dominant orientation angle of flow $\psi(p, q)$. In the present work, $G(m, n)$ and $\theta(m, n)$ are obtained from the Gabor magnitude and phase response. The projection of $G(m, n)$ on to the gradient orientation vector at (p, q) angle $\theta(p, q)$ is given by $G(m, n) \cos[\theta(m, n) - \theta(p, q)]$.

The sum-of-squares S of the projections of the Gabor magnitudes computed at each pixel of the neighborhood with the dominant orientation specified by Θ is given by

$$S = \sum_{m=1}^{P} \sum_{n=1}^{P} G^2(m, n) \cos^2[\theta(m, n) - \Theta]. \tag{7.14}$$

The sum varies as the reference orientation is varied, and attains its maximal value when Θ is perpendicular to the dominant orientation that represents the underlying texture in the given neighborhood [88, 89]. Differentiating S with respect to Θ yields

$$\frac{dS}{d\Theta} = 2 \sum_{m=1}^{P} \sum_{n=1}^{P} G^2(m, n) \cos[\theta(m, n) - \Theta] \sin[\theta(m, n) - \Theta]. \tag{7.15}$$

By setting $\frac{dS}{d\Theta} = 0$ and further simplifying the result, the solution for $\Theta = \Theta(p, q)$ that maximizes S at the point (p, q) in the image is given by [88]

$$\Theta(p, q) = \frac{1}{2} \tan^{-1} \left(\frac{\sum_{m=1}^{P} \sum_{n=1}^{P} G^2(m, n) \sin[2\theta(m, n)]}{\sum_{m=1}^{P} \sum_{n=1}^{P} G^2(m, n) \cos[2\theta(m, n)]} \right). \tag{7.16}$$

The second derivative, $\frac{d^2 S}{d\Theta^2}$, is given by

$$\frac{d^2 S}{d\Theta^2} = -2 \sum_{m=1}^{P} \sum_{n=1}^{P} G^2(m, n) \cos[2\theta(m, n) - 2\Theta]. \tag{7.17}$$

Equating Equation 7.17 to zero yields the estimated dominant angle of the texture $\psi(p, q)$ at (p, q) in the image to be

$$\psi(p, q) = \Theta(p, q) + \pi/2. \tag{7.18}$$

The coherence, $\gamma(p, q)$, at a pixel (p, q) is given by the cumulative sum of the projections of the Gabor magnitude responses for the pixels in a window of size $P \times P$, in the direction of the dominant orientation at the point (p, q) under consideration [88, 189], as

$$\gamma(p, q) = G(p, q) \frac{\sum_{m=1}^{P} \sum_{n=1}^{P} |G(m, n) \cos[\theta(m, n) - \psi(p, q)]|}{\sum_{m=1}^{P} \sum_{n=1}^{P} G(m, n)}; \tag{7.19}$$

in the present work $P = 15$, with (p, q) being located at the center of the $P \times P$ window.

7.5.2   ORIENTATION STRENGTH

In deriving the method for computing the orientation strength, it is assumed that rather than being single valued, the orientation information at a pixel is given by a function $G(\theta)$ that gives the strength of orientation (at the pixel location) in the underlying image for all angles θ. The measure of orientation strength at each pixel is computed, in the present work, as a weighted average of the Gabor magnitude responses, $G_k(m, n)$, for all directions of the filters used, $\theta_k, k = 0, 1, 2, \ldots, 179$, as

$$\alpha(m, n) = \sqrt{\frac{[\sum_{k=0}^{179} G_k(m, n) \cos(2\theta_k)]^2 + [\sum_{k=0}^{179} G_k(m, n) \sin(2\theta_k)]^2}{[\sum_{k=0}^{179} G_k(m, n)]^2}}. \tag{7.20}$$

This measure may also be termed "alignment energy." The results of application of the methods for computing the coherence and orientation strength are illustrated in Figures 7.12 and 7.13 for the mammographic images shown in Figures 4.5 (a) and 4.5 (b), respectively. It can be seen that the coherence images have appearances similar to that of the Gabor magnitude response images: both images have high responses for oriented structures [see Figure 4.6]. However, coherence represents an average measure of alignment over a neighborhood. On the other hand, the orientation strength takes into account the presence of multiple oriented structures intersecting at the pixel considered.

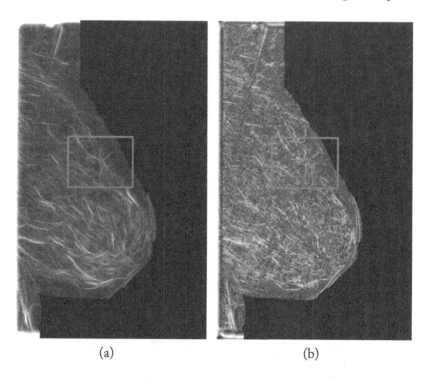

(a)   (b)

Figure 7.12: (a) Coherence and (b) orientation strength images for the mammogram shown in Figure 4.5 (a).

7.5.3   CHARACTERIZATION OF ANGULAR SPREAD

Three angular histograms or rose diagrams with 60 bins equally spaced over the angular range of $[-90°, 89°]$ were generated using the Gabor magnitude response, coherence, and orientation strength for each ROI based on the orientation field angle at each pixel. Another rose diagram was obtained using only the orientation field angle. To quantify the distribution based on the magnitude response, angle, coherence, and orientation strength, three entropy measures (Shannon's entropy, Tsallis entropy, and Rényi entropy) of each rose diagram, $F(\theta)$, were computed, after being normalized to have unit sum. The entropy measures are described in the subsequent sections. The angular spread of power in the frequency domain $S(\theta)$, as obtained by the procedure described in Section 7.3.4, is also included in the analysis.

Figures 7.14 and 7.15 illustrate the angular distributions (i.e., rose diagrams) obtained as above for the characterization of angular spread with a TP ROI and an FP ROI, respectively. TP ROIs

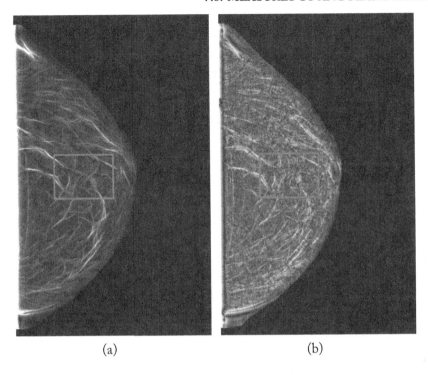

(a)   (b)

Figure 7.13: (a) Coherence and (b) orientation strength images for the mammogram shown in Figure 4.5 (b).

have oriented patterns in multiple directions, and the corresponding rose diagrams are expected to contain power in several angular bands. On the contrary, FP ROIs are not expected to have oriented patterns in multiple directions, and the related rose diagrams should contain power in fewer angular bands. The quantification of angular histograms is discussed in detail in the subsequent sections.

7.6   MEASURES OF ANGULAR DISPERSION

Entropy is a measure of information (in terms of order, disorder, or probability of occurrence) in a given set of data. Because texture can be considered a representation of the surfaces of objects, and surfaces are often composed of features with multiple orientations, analysis of the entropy of a given image may be a useful approach to the problem of texture classification [12, 39, 40, 42, 92, 190]. The most commonly used measure of order in a dynamical system is Shannon's or Boltzmann-Gibbs's entropy [191, 192]. Shannon's entropy is defined as

$$H_S = -\sum_i p_i \log_2 p_i, \tag{7.21}$$

where p_i is the probability of occurrence of an event i; the measure has its maximal value when all events are equally likely ($p_i = 1/N$, N is the number of events or bins). However, Shannon's entropy may not be able to characterize systems with long-range interactions, long-term memory effects, or abrupt changes. In this context, the use of higher order entropy, such as Tsallis entropy and Rényi entropy [192, 193], which are generalized forms of Boltzmann's or Gibbs's traditional entropy, could be used as alternatives of the typical entropy measures. Tsallis and Rényi entropies are both appropriate choices for a system with q-exponential behavior (an identity in the variable q that provides a known result in the limit, as $q \to 1$ for an exponential function) [42, 193].

7.6.1   TSALLIS ENTROPY

Tsallis entropy, a generalized form of Boltzmann-Gibbs's entropy, is defined as [193–195]

$$H_T(q) = \frac{1 - \sum_i p_i^q}{(q - 1)}, \tag{7.22}$$

where p_i is the probability of occurrence of an event i, and q is the moment order. Tsallis entropy is a generalized form of Shannon's traditional entropy, and is a nonextensive (scale-invariant) quantity for statistically independent subsystems. When $q \to 1$, Tsallis entropy recovers the definition of Shannon's entropy as follows [193, 194, 196]:

$$
\begin{aligned}
H_T(q) &= \frac{1 - \sum_i p_i^q}{(q - 1)} \\
&= \sum_i p_i \frac{p_i^{q-1} - 1}{1 - q} \\
&= \sum_i p_i \frac{2^{(q-1)\log_2 p_i} - 1}{1 - q} \\
&\approx \sum_i p_i \frac{[1 + (q - 1)\log_2 p_i] - 1}{1 - q} \\
&= -\sum_i p_i \log_2 p_i.
\end{aligned}
\tag{7.23}
$$

Tsallis entropy reaches its maximum when the probability of occurrence is the same for all events ($p_i = 1/N$, N is the number of events or bins); the maximum value of Tsallis entropy is given by [194–196],

$$H_{T\,\max} = \frac{N^{1-q} - 1}{1 - q}. \tag{7.24}$$

In order to quantify the angular spread of power, Tsallis entropy measures for the five angular distributions derived for each ROI were computed for various values of the order q. Based on an analysis of the classification performance using ROC analysis, Tsallis entropy of second order was used in the present work.

Tsallis entropy has found applications in physics, thermodynamics, and in wide areas of biomedical engineering [196–198]. It should be noted that the parameter q measures the degree of nonextensivity. For a mammographic image, it is not possible to characterize the underlying process as an extensive or a nonextensive system. Consequently, the entropy measures for nonextensive and extensive systems are applied using Tsallis and Rényi entropy.

7.6.2 RÉNYI ENTROPY

Rényi entropy [192, 193] is given by

$$H_R(q) = \frac{1}{(1-q)} \log_2 \left(\sum_i p_i^q \right), \tag{7.25}$$

where p_i is the probability of occurrence of an event i and q is the moment order. Rényi entropy is a generalized form of Shannon's traditional entropy, and is an extensive quantity for statistically independent subsystems, concave only for $0 < q < 1$. Rényi entropy recovers Shannon's entropy as a special case when $q \to 1$. Shannon's entropy is an averaged measure of information in the ordinary sense, whereas Rényi's measure represents an exponential mean over the same elementary information gains of $\log_2(1/p_i)$ [193].

It is worth mentioning that Rényi entropy has been widely used in multifractal theory [199], texture classification [200], pattern recognition, and image segmentation [201, 202].

To quantify the angular spread of power, Rényi entropy measures for the five angular distributions derived for each ROI were computed for various orders q. Based on an analysis of the classification performance using ROC analysis, Rényi entropy of eighth order was used in the present work; several works [199] have used $-10 \le q \le 10$.

Due to the presence of spiculations radiating at several angles, TP ROIs are expected to have a wide angular spread of power [40]. On the other hand, most FP ROIs were observed to contain a few intersecting ligaments, ducts, or vessels with the power limited to a small number of angular bands. Figures 7.14 and 7.15 illustrate the results of the procedures for the characterization of angular distribution with a TP ROI and an FP ROI, respectively. The values of the three entropy measures, i.e., Shannon's entropy (H_S), Tsallis entropy (H_T), and Rényi entropy (H_R) are also given in the captions.

7.7 ANALYSIS OF PERFORMANCE OF INDIVIDUAL FEATURES

Table 7.2 lists the A_z and p-values obtained for the 41 features used in the present study. The individual A_z and p-values of each of the features are presented for the full dataset of interval-cancer cases including normal control cases. The ranking of the features based on the A_z value is also

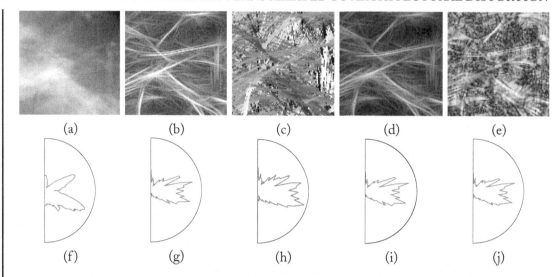

Figure 7.14: (a) A 128 × 128-pixel mammographic ROI with architectural distortion. (b) Gabor magnitude response. (c) Orientation field. (d) Coherence. (e) Orientation strength. (f) Angular histogram of power based on the Fourier spectrum of the image in (a). (g)–(j) Angular histograms of (b)–(e). Shannon's entropy (H_S), Tsallis entropy (H_T) of order two, and Rényi entropy (H_R) of order eight, respectively, are: (f) 7.2849, 0.9929, 6.6739; (g) 5.6168, 0.9765, 4.9968; (h) 5.6865, 0.9781, 5.1197; (i) 5.6002, 0.9761, 4.9699; and (j) 5.6192, 0.9765, 4.9882. Note: $H_{S\,max} = 7.49$, $H_{T\,max} = 0.9944$, $H_{R\,max} = 7.49$ for rose diagrams of type (f) with 180 bins and 5.91, 0.9833, and 5.91, respectively, for the types in (g)–(j) with 60 bins. The rose diagram in (f) has been rotated by $-90°$ to match the rose diagrams in (g)–(j). Reproduced with permission from: S. Banik, R. M. Rangayyan, and J. E. L. Desautels. "Rényi Entropy of Angular Spread for Detection of Architectural Distortion in Prior Mammograms." In *Proceedings of the 2011 IEEE International Symposium on Medical Measurements and Applications (MeMeA 2011)*, pp. 609–612, Bari, Italy, May 2011. © IEEE.

shown. The 4224 ROIs obtained from the full dataset of the interval-cancer cases and the normal control cases were studied for feature selection and classification; details of the performance of the selected features along with the relevant statistical analysis are presented in the next chapter.

With the node value, i.e., the results after the application of Gabor filters and phase-portrait analysis, the A_z obtained was 0.61. The highest AUC of 0.69 was obtained by the Rényi entropy measures of Gabor magnitude response (AS12) and coherence (AS14), indicating the stronger discriminating ability of angular spread of power (see Table 7.2) as compared to the other features. Most of the features related to the angular spread of power are ranked high, and most of Haralick's features have shown poor individual performance. In general, Laws' features have shown better individual performance as compared to Haralick's features.

However, none of the individual features has adequate discriminant power to classify the TP and FP ROIs efficiently; the results indicate the need for combinations of features and the application of feature selection. Regardless, node analysis serves as an important initial step to select candidate ROIs for further analysis.

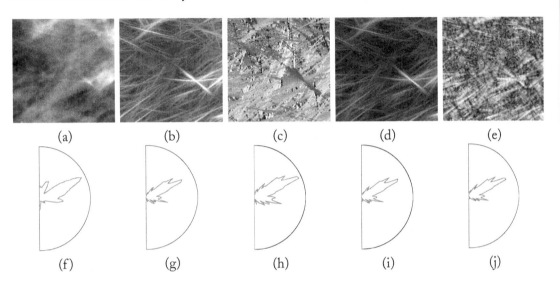

(a)   (b)   (c)   (d)   (e)

(f)   (g)   (h)   (i)   (j)

Figure 7.15: (a) A 128 × 128-pixel mammographic ROI; the ROI represents an FP detection due to overlapping or intersecting normal structures. (b) Gabor magnitude response. (c) Orientation field. (d) Coherence. (e) Orientation strength. (f) Angular histogram of power based on the Fourier spectrum of the image in (a). (g)–(j) Angular histograms of (b)–(e). Shannon's entropy (H_S), Tsallis entropy (H_T) of order two, and Rényi entropy (H_R) of order eight, respectively, are: (f) 7.1106, 0.9912, 6.1828; (g) 5.4097, 0.9706, 4.4368; (h) 5.4151, 0.9729, 4.5150; (i) 5.3901, 0.9701, 4.4111; and (j) 5.3603, 0.9693, 4.3605. Note: $H_{S\,max} = 7.49$, $H_{T\,max} = 0.9944$, $H_{R\,max} = 7.49$ for rose diagrams of type (f) with 180 bins, and 5.91, 0.9833, 5.91, respectively, for the types in (g)–(j) with 60 bins. The rose diagram in (f) has been rotated by $-90°$ to match the rose diagrams in (g)–(j). Reproduced with permission from: S. Banik, R. M. Rangayyan, and J. E. L. Desautels. "Rényi Entropy of Angular Spread for Detection of Architectural Distortion in Prior Mammograms." In *Proceedings of the 2011 IEEE International Symposium on Medical Measurements and Applications (MeMeA 2011)*, pp. 609–612, Bari, Italy, May 2011. © IEEE.

7.8   REMARKS

In this chapter, several methods for the characterization of texture patterns related to architectural distortion have been described. Further results of application of the techniques described in this chapter to mammograms are presented in the next chapter.

CHAPTER 8

Detection of Architectural Distortion in Prior Mammograms

In this chapter, the results obtained using the methods described in the previous chapters for the detection of sites of architectural distortion in prior mammograms of interval-cancer cases in a screening program are presented. It was hypothesized that screening mammograms obtained prior to the detection of breast cancer could contain subtle signs of early stages of breast cancer, in particular, architectural distortion. In this chapter, an analysis is presented of the results obtained for the detection of architectural distortion in prior mammograms of interval-cancer cases using Gabor filters, linear phase-portrait analysis, FD, Haralick's texture measures, Laws' texture energy measures, and measures of several types of entropy for the characterization of the angular spread of power with the magnitude and angle responses of Gabor filters, coherence, orientation strength, and power in the frequency domain.

8.1 ANALYSIS OF PERFORMANCE OF SELECTED FEATURES FROM VARIOUS SETS

Measures of the classification performance of the individual features are listed in Table 7.2. The ROC and FROC performance measures with various sets of selected features from the total set of 41 features using several classifiers are listed in Table 8.1. Feature selection was performed, in most of the cases, with stepwise logistic regression and the leave-one-image-out method for ROC and FROC analysis. In the 2-fold cross-validation method for ROC analysis, 50% of the TP ROIs and 50% of the FP ROIs were randomly selected for training, and the remaining ROIs were used for testing; the procedure was repeated 100 times. The Bayesian and SHL-ANN classifiers were used with the leave-one-image-out procedure for FROC analysis. The results of FROC analysis are reported for sensitivities of 0.8 and 0.9.

The node value and FD, on their own, are not adequate for accurate classification of the ROIs; A_z values of 0.62 and 0.61 were obtained with an SHL-ANN classifier for the node value and FD, respectively.

With reference to Table 7.2, the set LT1–LT10 consists of 10 features generated using Laws' texture energy measures. The best classification performance achieved with feature selection, in terms

Table 8.1: Results of ROC and FROC analysis using the selected features based on stepwise logistic regression for several types of feature sets. The dataset includes 106 prior mammograms of interval-cancer cases as well as 52 normal control mammograms. Feature selection and pattern classification were performed using the leave-one-image-out method; the mean and standard deviation values are presented for 100 trials in each case of the 2-fold cross-validated SHL-ANN. FD: fractal dimension; FLDA: Fisher linear discriminant analysis; Bayes: Bayesian classifier; ANN: artificial neural network; SHL: single-hidden-layer ANN; AS1–AS15: Features related to the angular spread of power; LT1–LT10: Laws' texture energy measures; HT1–HT14: Haralick's texture features.

Type of input features	Most frequently selected features	ROC analysis (A_z)				FROC analysis (FP/image at the sensitivities shown)			
		FLDA	Bayes	SHL	SHL, 2-fold	SHL		Bayes	
						80%	90%	80%	90%
Node		0.60	0.60	0.62	0.59 ± 0.02	10.6	14.5	10.2	15.6
FD		0.59	0.62	0.61	0.57 ± 0.03	16.8	18.6	12.4	15.6
LT1–LT10	LT1, LT7, LT6, LT2	0.70	0.71	0.71	0.71 ± 0.01	9.1	11.2	7.4	10.2
HT1–HT14	HT6, HT1, HT10, HT9	0.71	0.73	0.73	0.73 ± 0.02	9.4	12.6	6.7	9.5
AS1–AS15	AS12, AS13, AS1, AS10, AS3	0.73	0.73	0.74	0.75 ± 0.02	8.0	12.5	6.4	8.9
All	AS12, HT6, HT9, FD, HT1, Node, LT5, AS11	0.77	0.77	**0.79**	0.78 ± 0.03	7.0	10.0	**5.7**	8.8

of the A_z value, is 0.71 with the Bayesian and the SHL-ANN classifier; FROC analysis indicated a sensitivity of 0.8 at 7.4 FP/image. The most frequently selected features in this category (in the order of selection) are LT1, LT7, LT6, and LT2. In a related study, the selected features combined with the node value were found to provide a higher AUC value of 0.76 ± 0.01 with an ANN-RBF classifier and a sensitivity of 0.84 was obtained at 9.5 FP/image using the SHL-ANN classifier [87].

Haralick's 14 texture features have shown performance comparable to that of Laws' texture energy measures. The most frequently selected features in this category are HT6, HT1, HT10, and HT9; a sensitivity of 0.8 was obtained at 6.7 FP/image with the Bayesian classifier. In a related work, Haralick's texture features combined with the node value and FD resulted in AUC values of 0.77 ± 0.01 with an ANN-RBF classifier and 0.77 ± 0.03 with an SVM classifier [39]. FROC analysis indicated a sensitivity of 0.80 at 7.6 FP/image using the SHL-ANN classifier [39].

Using the 15 features related to the measures of angular spread (AS1–AS15), the most frequently selected features were found to be AS12, AS13, AS1, AS10, and AS3. This selection represents a combination of all three types of entropy measures used, i.e., Shannon's entropy, Tsallis entropy, and Rényi entropy. The selected features using stepwise logistic regression and the leave-one-image-out method resulted in the A_z value of 0.74 with an SHL-ANN classifier; a sensitivity of 0.8 was obtained at 6.4 FP/image with the Bayesian classifier. The results indicate that the measures of angular spread have a strong capability of characterizing the oriented texture patterns related to architectural distortion. Although the results obtained are comparable to those provided by Laws' and Haralick's texture measures (with marginal improvement in some cases), the measures are more meaningful and related to the expected oriented patterns.

In a related study, node analysis, fractal analysis, and measures of angular spread of power in the frequency domain resulted in a sensitivity of 0.82 at 7.7 FP/image using the same dataset [84]; an A_z value of 0.73 ± 0.02 was obtained using an ANN-RBF classifier with the 2-fold random subsampling and cross-validation method. In another related work [188], the node value and the measures for characterization of the angular spread of power were used. For characterization of the angular spread, each ROI was represented by the Shannon's entropy of the angular histogram composed with the Gabor magnitude response, angle, coherence, and orientation strength; the entropy of the angular spread of power in the Fourier spectrum was also used. A_z values of 0.75 and 0.76 were obtained using SHL-ANN and ANN-RBF, respectively; FROC analysis indicated 82% sensitivity at 7.2 FP/image. Similar studies were conducted separately on the same dataset using Tsallis entropy [187] and Rényi entropy [203] for characterization of the angular spread of power: A_z values of 0.74 and 0.75 were obtained, respectively, using an ANN-RBF. FROC analysis indicated a sensitivity of 0.80 at 7.1 FP/image using the leave-one-image-out method with an SHL-ANN classifier for both the cases [187, 203].

The final set includes all of the 41 features in the present study. The results achieved, in terms of the A_z value, using the features selected by stepwise logistic regression and the leave-one-image-out method, are 0.77 with FLDA and the Bayesian classifier, and 0.79 with the SHL-ANN classifier. FROC analysis of the performance of the features resulted in sensitivities of 0.80 and 0.90 at 5.7 and 8.8 FP/image, respectively, with the Bayesian classifier and the leave-one-image-out method. Sensitivies of 0.80 and 0.90 were achieved at 7.0 and 10.0 FP/image with the SHL-ANN calssifier. The FROC curves obtained using the SHL-ANN and the Bayesian classifiers with the features selected by stepwise logistic regression are shown in Figure 8.1.

The results obtained, using all of the features studied, in terms of A_z values, show improvement over the results obtained using the individual types or groups of features. Although the combination of all types of features did not generate significantly improved FROC results, combinations of two or more types of features were found to provide improved classification accuracy in a related study [40].

The reduction of FPs in the final result is shown in Figure 8.2 for the mammograms presented in parts (a) and (b) of Figure 5.3. The results obtained using stepwise logistic regression and the Bayesian classifier with the leave-one-image-out method are presented; only the top six ROIs based on the discriminant values are shown for each image, with the corresponding average sensitivity of 0.8 at 5.7 FP/image. The numbers outside the parentheses represent the ranking based on the discriminant values obtained by the Bayesian classifier and the numbers within the parentheses represent the earlier ranking based on the node value. From Figure 8.2, it is evident that the use

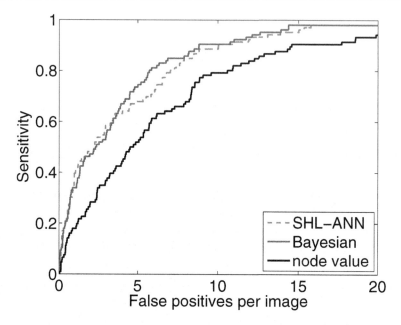

Figure 8.1: FROC curves for the dataset of 106 prior mammograms of interval-cancer cases and 52 normal mammograms with the selected features using stepwise logistic regression and the SHL-ANN and the Bayesian classifiers with the leave-one-image-out method. The FROC curve generated using the node value only (without any classifier) is also shown for reference. SHL-ANN: single-hidden-layer backpropagation artificial neural network; sensitivity = true-positive fraction.

of the additional features has led to a substantial reduction of FPs in the detection of architectural distortion. For the mammogram shown in part (a) of Figure 8.2, two TP ROIs were included within the top six ROIs based on the discriminant values obtained using the Bayesian classifier. Only one TP ROI was included within the top six ROIs for the image shown in part (b) of Figure 8.2.

Figure 8.3 (a) shows a prior mammogram with a subtle appearance of architectural distortion; the corresponding ROIs detected are also shown: the site of architectural distortion has been captured by the 24th peak in the node map. No TP ROI was included in the top six ROIs based on the discriminant values as illustrated in part (b) of Figure 8.3. Due to the subtle appearance of the TP ROI, it was ranked 24 in the node map, and the methods used failed to include the corresponding TP ROI within the top six ROIs based on the discriminant values obtained using the Bayesian classifier. However, even if only the top six ROIs are considered, the average sensitivity remains above 80% over the entire dataset including the normal control cases.

The FROC curves using the Bayesian classifier with the features selected by stepwise logistic regression are shown in Figure 8.4 for all the sets of features listed in Table 8.1. The FROC curves obtained using each set of features and all combined features indicate a substantial reduction of the

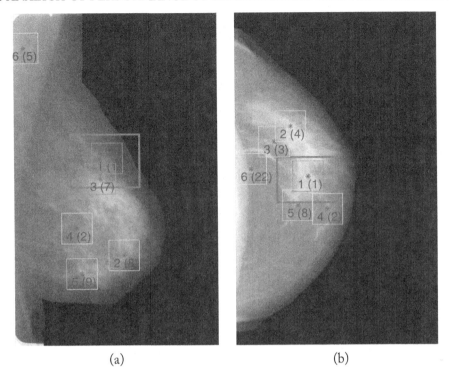

(a)   (b)

Figure 8.2: Results obtained using the selected features from all of the 41 features and the Bayesian classifier with the leave-one-image-out method; only the top six ROIs based on the discriminant values are shown for each image, with the associated average sensitivity of 0.8 at 5.7 FP/image. (a) Reduction of FPs for the image shown in Figure 5.3 (a). (b) Reduction of FPs for the image shown in Figure 5.3 (b). The numbers outside the parentheses represent the ranking based on the discriminant values obtained by the Bayesian classifier, and the numbers within the parentheses represent the earlier ranking based on the node value.

number of FPs per image in the detection of architectural distortion compared to the initial stage of the study, that is, node analysis. Haralick's texture features, Laws' energy measures, and the measures of angular spread produced similar FROC curves; thus, the results are comparable and could be replaced by one another.

8.1.1   STATISTICAL SIGNIFICANCE OF DIFFERENCES IN PERFORMANCE

The results obtained using all types of the features studied are slightly better than those obtained using individual feature sets; the results of analysis of the statistical significance of the differences between the various ROC and FROC curves obtained are presented in this section. Because the SHL-ANN classifier was found to perform well in ROC analysis and the Bayesian classifier performed well in

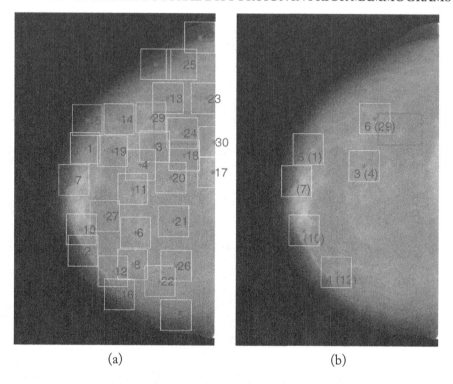

(a)   (b)

Figure 8.3: (a) The 30 ROIs obtained automatically using the peaks detected in the node map for the prior mammogram of an interval-cancer case. (b) Results obtained using the selected features from all of the 41 features and the Bayesian classifier with the leave-one-image-out method; only the top six ROIs based on the discriminant values are shown, with the associated average sensitivity of 0.8 at 5.7 FP/image. The numbers outside the parentheses represent the ranking based on the discriminant values obtained by the Bayesian classifier and the numbers within the parentheses represent the earlier ranking based on the node value.

FROC analysis, the results of the SHL-ANN classifier and the Bayesian classifier were used for the analysis of the statistical significance of the differences between the ROC curves and FROC curves, respectively.

Table 8.2 gives the measures of the statistical significance of the differences between the ROC curves generated using the SHL-ANN classifier and the leave-one-image-out method for the four sets of selected features as well as the node value and FD, as shown in Table 8.1. The results obtained using the selected features from the combination of all the features show statistically highly significant improvement over the results obtained with the node value, FD, and each of the three individual feature sets. The ROC curves of the node value and FD are not statistically significantly different; the ROC curve of the set of measures of angular spread is statistically significantly different

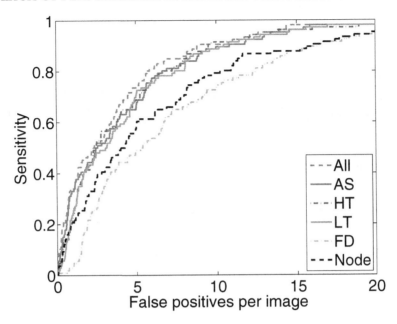

Figure 8.4: FROC curves for the dataset of 106 prior mammograms of interval-cancer cases and 52 normal control mammograms with the selected features from several sets of features using stepwise logistic regression and the Bayesian classifier with the leave-one-image-out method.

from that of Haralick's texture measures, and different from the remaining three with high statistical significance.

The results demonstrate that the new measures of angular spread developed in the present study can provide statistically significant improvement in classification performance than the other sets of features studied, as determined by ROC analysis.

Table 8.3 illustrates the statistical significance of the differences in FROC results, using the Bayesian classifier and the leave-one-image-out method for the four sets of selected features as well as the node value and FD, obtained by comparing the discriminant values using the JAFROC1 software [21, 150, 151]. The discriminant values obtained for FROC analysis with the node value and FD are different from each other and each of the other selected feature sets with high statistical significance. The differences between the FROC curves for the sets of the discriminant values obtained for the other feature sets are not statistically significant; therefore, any one of the sets of features LT1–LT10, HT1–HT14, or AS1–AS15 may be used. Although the results obtained using the selected features from the different feature sets indicate no statistically significant improvement as compared to one another, it is evident from all of the results listed in Table 8.1 that the combination of all of the 41 features studied in the present work could be used to facilitate efficient detection of architectural distortion in prior mammograms.

Table 8.2: Analysis of the statistical significance, using the p-value, of the differences between the ROC curves obtained using the SHL-ANN and the leave-one-image-out method for the selected features from several types of feature sets. The p-values were estimated using ROCKIT (up to four decimal places).

Feature set.	FD	LT1–LT10	HT1–HT14	AS1–AS15	All
Node	0.1109	0.0000	0.0000	0.0000	0.0000
FD		0.0000	0.0000	0.0000	0.0000
LT1–LT10			0.0006	0.0000	0.0000
HT1–HT14				0.0213	0.0000
AS1–AS15					0.0001

Table 8.3: Analysis of the statistical significance, using the p-value, of the differences between the FROC curves obtained using the Bayesian classifier and the leave-one-image-out method for the selected features with several feature sets. The p-values were estimated using JAFROC1 (shown up to four decimal places).

Feature set.	FD	LT1–LT10	HT1–HT14	AS1–AS15	All
Node	0.0038	0.0000	0.0000	0.0000	0.0000
FD		0.0000	0.0000	0.0000	0.0000
LT1–LT10			0.1893	0.2187	0.0954
HT1–HT14				0.1542	0.2895
AS1–AS15					0.5136

8.1.2 EFFECTS OF VARIOUS TYPES OF CROSS-VALIDATION AND TRAINING SETS

If the feature selection step is performed using the entire dataset of interval-cancer cases including normal control cases, a positive bias may be introduced in the results obtained by the LOO method or by subsampling and cross-validation methods. To study this effect on the results of the present work, feature selection and pattern classification were performed using several methods to separate the training and testing sets; in particular, the LOO and the 2-fold random subsampling cross-validation methods were applied on the basis of patients, images, and ROIs. In the 2-fold random subsampling cross-validation procedure, approximately 50% of the normal and 50% of the abnormal or TP ROIs, images, or patients were randomly selected for the training procedure (including feature selection) and the rest were used in the testing step, and the procedure was repeated 100 times.

Table 8.4 shows the results obtained using the selected features from all of the sets of features. It should be noted that different sets of features could be selected in each of the trials depending on the characteristics of the randomly selected ROIs, images, or patients in the training set. See Section 8.1.3 for more discussion on this topic. The results obtained using the various cross-validation methods, with the features selected using the entire dataset or using the training set only in each trial, indicate that no significant bias is found to be present in the results; the corresponding p-values are shown in Table 8.5. The p-values show no statistical significance in the results obtained using the different cross-validation methods.

Table 8.4: Results of ROC analysis using the selected features based on stepwise logistic regression for several cross-validation methods and the SHL-ANN classifier. SHL: single-hidden-layer ANN; LOO: leave-one-out method.

Cross-validation method	ROC analysis (A_z)	
	Feature selection using the training set	Feature selection using the entire dataset
LOO-ROI	0.80	0.80
LOO-image	0.79	0.79
LOO-patient	0.78	0.78
2-fold, ROI	0.78 ± 0.03	0.78 ± 0.02
2-fold, image	0.77 ± 0.03	0.77 ± 0.02
2-fold, patient	0.78 ± 0.02	0.78 ± 0.02

Table 8.5: Analysis of the statistical significance (p-value) of the differences between the ROC curves obtained using several cross-validation methods and the SHL-ANN classifier with the selected features from the set of all of the 41 features studied. The selected features were obtained using the entire dataset (indicated by [1]) or using the training set only (indicated by [2]). The p-values were estimated using ROCKIT (up to four decimal places). LOO: leave-one-out; ROI: region of interest.

Method	LOO-image[1]	LOO-image[2]	LOO-patient[1]	LOO-patient[2]
LOO-ROI[1]	0.5327	0.7334	0.4172	0.4795
LOO-image[1]		0.7739	0.2362	0.3251
LOO-image[2]			0.1938	0.2760
LOO-patient[1]				0.2881

8.1.3 ANALYSIS OF PERFORMANCE WITH VARIOUS METHODS OF CROSS-VALIDATION AND FEATURE SELECTION

The performance of the methods used was evaluated with several cross-validation techniques in feature selection and pattern classification. Table 8.6 illustrates the results obtained with the full set of 41 features using two feature selection methods: stepwise logistic regression and stepwise regression; with the LOO-based cross-validation method, feature selection and training of the classifier were performed on the training set (all but one), and tested on the remaining ROI, image, or patient/individual. Table 8.6 demonstrates that the stepwise logistic regression and stepwise regression methods for feature selection yield similar results although different features could be selected by the two different feature selection methods. The different cross-validation methods also produced similar results.

Table 8.6: Results of ROC analysis using the selected features based on stepwise logistic regression or stepwise regression for several cross-validation methods and for several classifiers. ANN: artificial neural network; LOO: leave-one-out method; SHL: single-hidden-layer ANN; FLDA: Fisher linear discriminant analysis. The mean and standard deviation values are presented for 100 trials in each case of the 2-fold random subsampling and cross-validation method.

Cross-validation method	Stepwise logistic regression			Stepwise regression		
	FLDA	Bayesian	SHL-ANN	FLDA	Bayesian	SHL-ANN
LOO-ROI	0.77	0.77	0.80	0.78	0.77	0.79
LOO-image	0.77	0.77	0.79	0.77	0.77	0.78
LOO-patient	0.76	0.76	0.78	0.77	0.76	0.78
ROI, 2-fold	0.75 ± 0.01	0.76 ± 0.02	0.78 ± 0.03	0.75 ± 0.02	0.75 ± 0.01	0.78 ± 0.02
Image, 2-fold	0.77 ± 0.01	0.77 ± 0.02	0.77 ± 0.03	0.76 ± 0.01	0.76 ± 0.02	0.78 ± 0.01
Patient, 2-fold	0.75 ± 0.02	0.76 ± 0.01	0.78 ± 0.02	0.75 ± 0.02	0.75 ± 0.02	0.78 ± 0.02

Figure 8.5 illustrates the histograms of the selected features using stepwise logistic regression and stepwise regression. The set of features selected using stepwise logistic regression includes all types of features, with three of Haralick's texture measures. Although Haralick's textures features were ranked low (see Table 7.2) in terms of individual A_z values, they were selected by the feature selection method. This indicates that the features with high A_z values may not provide improved results when combined with other high-ranking features. On the other hand, certain features with lower discriminating ability on their own may add useful information when combined with other features and improve the overall classification accuracy. The features often selected with stepwise regression also include all types of features; however, in this case, the features related to angular spread are the most prominent type (with more of them selected than others) and Laws' energy measures are the least prominent type (with only two of them selected only a few times).

The FROC curves shown in Figure 8.6 were obtained using the two feature selection methods discussed above with the leave-one-image-out method and the Bayesian classifier. The results indicate sensitivities of 0.80 and 0.90 at 5.7 and 8.8 FP/image, respectively, for the features selected

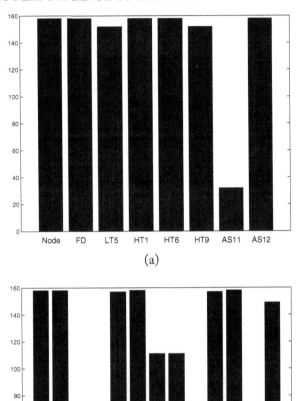

(a)

(b)

Figure 8.5: Histograms of the selected features using two feature selection methods with the leave-one-image-out procedure. Features selected by: (a) stepwise logistic regression, and (b) stepwise regression based on the p-value of the F-statistic. FD: fractal dimension; AS1–AS15: Features related to the angular spread of power; LT1–LT10: Laws' texture energy measures; HT1–HT14: Haralick's texture features. The ordinate represents the number of times a given feature was selected in a total of 158 trials. The features not listed along the abscissa were not selected even once among the 158 trials; see Table 7.2 for the full list of features derived in the present study.

by stepwise logistic regression, and at 5.2 and 9.8 FP/image, respectively, for the features selected by stepwise regression. The FROC data obtained with the features selected using the two feature selection methods showed no statistically significant difference ($p = 0.1539$). Thus, it can be concluded that the use of a reasonable method for feature selection and a trained classifier could significantly improve the results from the initial detection, i.e., the node value.

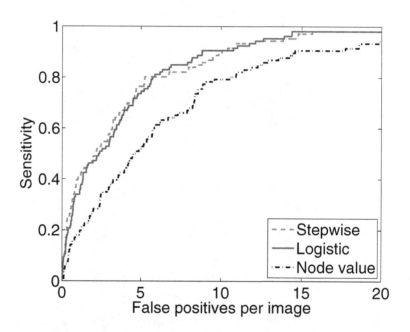

Figure 8.6: FROC curves for the dataset of 106 prior mammograms of interval-cancer cases and 52 normal control mammograms obtained using two feature selection methods with the leave-one-image-out procedure and the Bayesian classifier. The FROC curve generated using the node value only (without any classifier) is also shown for reference.

8.2   ANALYSIS OF PERFORMANCE WITH SUBSETS OF THE INTERVAL-CANCER DATASET

According to the indication given by the radiologist (J.E.L.D.), 38 of the 106 prior mammographic images had visible architectural distortion and the remaining 68 images had questionable or no clearly evident architectural distortion. In this context, the feature selection and pattern classification techniques were applied to evaluate the performance on each of the subsets of 38 and 68 images. Table 8.7 summarizes several aspects of the subsets partitioned based on the images with clearly evident architectural distortion, images with no evident architectural distortion, images of the interval-cancer cases only (i.e., the combination of the previous two subsets), and images of all interval-cancer cases with normal cases.

Table 8.7: Summary of the dataset of interval-cancer cases and organization of the subsets.

Number of	Evident	Not evident	Interval cancer	Interval cancer and normal
Patients/individuals	21	35	56	69
Images	38	68	106	158
ROIs	1038	1783	2821	4224
TP-ROIs	133	168	301	301
FP-ROIs	905	1615	2520	3923
Average ROIs/image	27.3	26.2	26.6	26.7
Average *TP*/image	3.5	2.5	2.8	1.9
Average *FP*/image	23.8	23.8	23.8	24.8

From Table 8.7, it can be seen that, for the clearly evident architectural distortion cases, the average numbers of ROIs/image and TP-ROIs/image are higher as compared to those for the other subsets. Although each image with architectural distortion contains a single site of architectural distortion, some scattered or separate parts of the architectural distortion pattern present in the image could be identified in the node analysis step. Table 8.8 summarizes the results of ROC and FROC analysis obtained; for feature selection, stepwise logistic regression with the leave-one-image-out method was used; classification was performed using the leave-one-image-out technique with several classifiers. Despite the high number of TP-ROIs/image, ROC analysis indicates lower classification accuracy for the clearly evident cases of architectural distortion. The clearly evident cases have 38 images with 1038 ROIs and 41 features. As a consequence, the training set for feature selection and pattern classification is small for accurate representation of the true classes, and the results obtained could be pessimistically biased [152, 204]. On the other hand, the images with no evident architectural distortion have led to results comparable with those obtained using the full interval-cancer dataset. The study with the images of all interval-cancer cases excluding the normal cases has provided improved results. The inclusion of normal cases did not substantially degrade the performance of the methods.

It should be noted that different feature sets could get selected depending on the size and characteristics of the datasets, and that the number of images or ROIs in the training set and their characteristics play important roles in pattern classification. The design of classifiers that can accurately distinguish abnormal from normal features is an important and critical step in the development

Table 8.8: Results of ROC and FROC analysis with subsets of images of interval-cancer cases and normal cases. For feature selection, stepwise logistic regression with the leave-one-image-out method was used; classification was performed using the leave-one-image-out technique with several classifiers. ANN: artificial neural network; SHL: single-hidden-layer ANN; FLDA: Fisher-linear discriminant analysis. For ROC analysis with individual features, the feature with the corresponding A_z value is indicated in parentheses.

Analysis Method	Classifier	Evident	Not evident	Interval cancer	Interval cancer and normal
ROC analysis with individuals	maximum	0.68 (AS7)	0.68 (AS12)	0.68 (AS12)	0.69 (AS12)
features (A_z)	minimum	0.51 (HT2)	0.50 (HT2)	0.51 (HT13)	0.50 (HT14)
ROC analysis	FLDA	0.72	0.77	0.76	0.77
with selected	Bayesian	0.70	0.76	0.77	0.77
features (A_z)	SHL-ANN	0.72	0.77	0.78	0.79
FROC analysis	Bayesian 80%	6.6	7.1	5.9	5.7
(FP/image	Bayesian 90%	13.2	10.2	8.4	8.8
at the	SHL-ANN, 80%	5.7	6.5	6.0	7.0
sensitivities shown)	SHL-ANN, 90%	9.3	8.5	9.2	10.0

of CAD algorithms [204]. It has been shown that the performance of a classifier with unknown cases depends on the sample size used in the training step [152]. When a finite training set size is used, the performance could be pessimistically biased in comparison to that obtained from an infinitely large training set. In order to design a classifier with a performance generalizable to the population at large, one has to use a sufficient number of case samples that are representative of the population, which is demonstrated in the present study.

8.3 CROSS-VALIDATION OF PERFORMANCE IN DETECTION USING DIFFERENT DATASETS

To test the effects of the training set on the results, the methods were applied to the second dataset of 14 prior mammograms and 14 detection mammograms of seven screen-detected cancer cases; see Table 5.2. As discussed in Section 5.2.2, Gabor filters and phase-portrait analysis did not detect any site of architectural distortion in three of the prior mammograms and one of the detection mammograms in the cases of screen-detected cancer; these mammograms had no clearly evident architectural distortion as indicated by the radiologist. Therefore, the maximum sensitivity achieved after the initial detection was 79% and 93% with the prior and detection mammograms, respectively. Table 8.9 summarizes the results of training and testing with different combinations of the two datasets using the feature selection and pattern classification methods applied with all of the 41 features derived.

The results shown in Table 8.9 indicate that the use of prior mammograms for training and the use of detection mammograms or mammograms of interval-cancer cases including normal control cases for testing produces better results than using the detection mammograms for training. The results also demonstrate the need for training the classifier with the prior mammograms of interval-cancer cases to obtain good classification accuracy in the detection of architectural distortion in prior mammograms of interval-cancer cases.

It should be noted that the dataset of the screen-detected cancer cases is small, and the use of the prior or detection mammograms in the training process may not provide a good capability of generalization of the classifier. On the contrary, the use of the larger dataset of interval-cancer cases could provide improved results in the classification of the smaller datasets of screen-detected cases. The size of the training set, relative to the number of features used in pattern classification, affects the accuracy and reliability of the results [142, 165, 168]. The number of features should not be increased without proportionately increasing the number of training samples; for a fixed number of training samples, the addition of more features could lead to poorer performance, termed as the "curse of dimensionality." The process may lead to overtraining, resulting in the lack of generalization for other datasets. Similar effects could exist in the present study. Although it is desirable to find the optimal number of features to be used in a pattern recognition process for a particular size of dataset [165], the various combinations of features could constrain reaching such a decision, and exhaustive combinations of features may need to be evaluated. Nonetheless, the observations arising from the present study of cross-validation agree with those of other similar studies reported in the literature [95].

8.4 COMPARATIVE ANALYSIS AND DISCUSSION

The detection of architectural distortion in prior mammograms is a difficult problem. In the present study, Gabor filters, linear phase-portrait analysis, fractal analysis, analysis of the angular spread of power, structured pattern analysis using Laws' texture energy measures, and Haralick's methods

Table 8.9: Results of ROC and FROC analysis using the selected features from the set of all the 41 features derived based on stepwise logistic regression for several combinations of training and testing datasets. Bayes: Bayesian classifier; SHL: single-hidden-layer ANN; LOO: leave-one-out method; SDC: screen-detected cancer; Interval: prior mammograms of interval-cancer cases.

Training set	Images	Test set	Images	ROC (A_z) Bayes	SHL	FROC analysis (Bayesian classifier)
Detection	14	Prior of SDC	14	0.71	0.81	79% sensitivity at 19.0 FP/image
Prior of SDC	14	Detection	14	0.82	0.82	79% and 93% sensitivity at 7.3 and 18.4 FP/image
Detection	14	Interval & Normal	158	0.69	0.71	80% and 90% sensitivity at 9.2 and 12.6 FP/image
Interval & Normal	158	Detection	14	0.82	0.82	79% and 93% sensitivity at 7.6 and 20.1 FP/image
Prior of SDC	14	Interval & Normal	158	0.71	0.75	80% and 90% sensitivity at 7.5 and 11.2 FP/image
Interval & Normal	158	Prior of SDC	14	0.77	0.83	79% sensitivity at 16.6 FP/image
Interval & Normal	158	Prior of SDC & Detection	28	0.79	0.81	79% sensitivity at 15.9 FP/image
Prior of SDC & Detection	28	Interval & Normal	158	0.72	0.77	80% and 90% sensitivity at 7.8 and 11.8 FP/image
Detection	–	LOO method	–	0.81	0.80	79% and 93% sensitivity at 8.5 and 16.6 FP/image
Prior of SDC	–	LOO method	–	0.72	0.71	79% sensitivity at 12.6 FP/image
Interval & Normal	–	LOO method	–	0.77	0.79	80% and 90% sensitivity at 5.7 and 8.8 FP/image

for statistical analysis of texture were used for the detection of architectural distortion in prior mammograms of interval-cancer cases.

The estimation of FD using the Fourier power spectrum method is widely used for texture analysis and pattern classification; however, in the estimation of FD, the distribution of power over radial frequency is considered and the information regarding the angular spread of power is overlooked. In the present study, the angular spread of power in the Fourier spectrum is analyzed via geometric transformation to polar coordinates and used to generate discriminative features for the detection of radiating spiculated patterns related to architectural distortion. Although the Hankel transform [173] and the Fourier transform in polar coordinates have been used for texture analysis and pattern recognition [174], the method developed in the present work to characterize the angular spread of power is novel and is found to be effective for the analysis of oriented texture.

Haralick's texture measures are based upon statistical properties of the pixels in the given image, and some of Haralick's features are difficult to relate to the structural characteristics of an image; they may not be directly related to the intersecting structures, spiculations, and node-like patterns of architectural distortion. In this context, Laws' texture energy measures, being based on convolution kernels that emphasize specific structural patterns, can complement by addressing some of the limitations of Haralick's texture measures. When used together, they form a powerful tool to characterize or analyze texture and oriented patterns in a given image [40]. The geometric transformation of the ROI being processed so as to convert spiculating or radiating patterns to waves or ripples is an important prerequisite to the derivation of Laws' texture energy measures for the characterization of architectural distortion, and represents an important and novel contribution of the present work.

In a related study, Rangayyan et al. [39] reported a sensitivity of 0.8 at 7.6 FP/image using linear phase-portrait analysis, FD, and Haralick's texture features. In the present study, using the selected features from the full set of features has resulted in an encouraging sensitivity of 0.9 at 8.8 FP/image. The proposed methods have demonstrated the ability to detect early signs of breast cancer 15 months ahead of the time of clinical diagnosis, on the average, for interval-cancer cases, with a sensitivity of 0.8 at 5.7 FP/image. Column 2 in Table 8.1 lists the features selected from each set; although such sets of features obtained by feature selection techniques vary depending upon the selection technique or criteria used and the characteristics of the image dataset, detailed analysis of the selected features and the related ROC or FROC performance could assist in the design of a practical tool.

Characterization of the angular spread of power in the frequency domain has been shown to be effective in the detection of architectural distortion in a related previous study [84]. In this context, analysis of the angular distribution of the magnitude and angle responses of Gabor filters, coherence, orientation strength could add important information regarding the spiculating patterns of architectural distortion. Due to the presence of spicules radiating at several angles, TP ROIs are expected to have a wide angular spread of power, whereas most FP ROIs should contain only a few intersecting ligaments, ducts, or vessels with the power limited to a small number of angular bands. The entropy measures presented here for the characterization of angular distribution have shown good performance in the detection of architectural distortion [42, 43, 187, 188, 190, 203]. The use of the higher order entropy measures, such as Tsallis and Rényi entropy, along with the conventional Shannon's entropy, for characterization of the angular spread has led to a small reduction of FPs in the detection of architectural distortion. The measures of angular spread have produced better results in terms of ROC and FROC analysis than the other types of features derived in the present study. When combined with other types of features, they could provide complementary information regarding the oriented and/or spiculating texture patterns. In addition, they could be effectively used instead of Haralick's texture features or Laws' texture energy measures because of the more obvious relation with the structure of the associated oriented texture patterns. However, in the present work, all the pixels within a given ROI contribute to the angular distribution; improved discriminating ability could be achieved by considering the corresponding CLS pixels only. Selecting the appropriate orders for Tsallis entropy and Rényi entropy is challenging; in the present study, the orders were selected based on experimentation. Selection of the appropriate range of frequencies for characterization of angular spread in the frequency domain is also critical and could affect the result; care should be taken in selection of the proper frequency range.

In the context of the results of the previous works on architectural distortion [29, 38, 39, 54, 55, 58, 70, 73, 77–79], the results obtained in the present work with prior mammograms of interval-cancer cases, including normal control cases, are comparable, encouraging, and better in many aspects; see Table 2.1. Nakayama et al. [78] performed multiresolution analysis and obtained a sensitivity of 71.3% (57 out of 80 images) at 3.01 FP/image. Nemoto et al. [29] proposed a method to detect architectural distortion with radiating spiculation on 25 digital mammograms, and obtained a sensitivity of 80.0% at 0.80 FP/image; however, the dataset was small and included evident architectural distortion with radiating spiculation, and did not include any normal control cases. The automatic detection of architectural distortion in prior mammograms of interval-cancer cases, which is a more difficult problem due to the subtle and ill-defined appearance of the associated patterns, has not been studied adequately. Detailed comparative analysis is not possible because of the variability of the size and types of the datasets used in the various related works. It should be noted that the commonly used public databases of mammograms, such as the DDSM [80] and the MIAS database [14], do not contain prior mammograms.

Simultaneous analysis of current and prior mammograms is usually recommended by radiologists for use in the screening and diagnosis of breast cancer: the same approach could be used to enhance the performance of CAD systems. Sameti et al. [41] reported an average classification rate of 72% using six selected texture and photometric features computed from manually marked regions on the last screening mammograms prior to the detection of breast cancer (see Table 2.2 for a list of related studies on prior mammograms). In a related work, Rangayyan et al. [38] used phase portraits, FD, and Haralick's texture features for the detection of architectural distortion in prior mammograms of screen-detected cancer, and achieved a sensitivity of 79% at 8.4 FP/image with a set of 14 prior mammograms. The dataset was small, with only 14 prior mammograms and 14 detection mammograms, and with no normal control cases as used for cross-validation in the present study and shown in Table 8.9; the detection mammograms were available for reference to the radiologist when labeling the areas with architectural distortion on the prior mammograms.

In the analysis with the LOO and the 2-fold random subsampling cross-validation methods based on ROIs, it is assumed that all ROIs are independent. It has been suggested that the TP ROIs from the same case could be correlated [205]. To address this concern, the LOO and the

2-fold random subsampling cross-validation methods were also applied on the basis of patients and images; see Section 8.1.2. Furthermore, the radiologist (J.E.L.D.) indicated that, in the present work on the detection of architectural distortion, TP ROIs originating from the CC and MLO views of the same breast might not be correlated; architectural distortion could be evident in one view and the other view of the same breast may not contain any distinct sign of architectural distortion. In addition, even if TP ROIs are present in both views, their textural characteristics and appearance could be different.

One of the shortcomings in the present study is that the uncertainty related to the FROC curves, the number of FPs reported at a given sensitivity, and several of the A_z measures derived has not been analyzed. The confidence bounds of the corresponding ROC and FROC curves can be obtained through resampling methods including the bootstrap, jackknife, and permutation tests [206–208]. Although ROCKIT and JAFROC can be used to obtain confidence bounds for ROC and FROC curves, respectively, they have not been derived in the present work.

In the present study on interval-cancer cases, including normal control cases, the diagnostic mammograms were not available to aid the process of localization of the sites of architectural distortion in the prior mammograms. The radiologist (J.E.L.D.) who analyzed and annotated the images used in the present study has more than 40 years of experience in mammography, of which more than 20 years is in screening for breast cancer; he was also a member of the team of radiologists in the Screen Test Program and interpreted the mammograms at the original instances of screening. All cases of interval cancer were reviewed by a panel of five experienced radiologists in the screening program as part of the standard protocol. The laterality, location, and nature of architectural distortion and/or other signs of breast cancer were determined by the radiologists and pathologists involved in the diagnostic imaging and other investigations; whereas the detection or diagnostic mammograms were not available for the present study, the associated reports were available for some of the cases. These issues may have affected the size and positional accuracy of the suspicious areas labeled by the radiologist.

In the present work, 68 of the 106 prior mammographic images of the interval-cancer cases contained questionable or no clearly evident architectural distortion (as determined by the radiologist). In addition to the subtle and hard-to-detect appearance of the architectural distortion present in the prior mammograms, some of the automatically detected TP ROIs were ranked low in the step of node map analysis; regardless, they were labeled as TP ROIs if their centers were found to be within the related area delineated by the radiologist (see Figures 5.3 and 8.3). Some of the TP ROIs may contain only a part or small portions of the spicules originating from the focal point of architectural distortion; such ROIs could increase the ambiguity and difficulty in pattern classification. These factors affect the feature selection and classification process, and could be underlying causes of the lower A_z values obtained for the individual features as compared to the typical A_z values in studies on the detection of more readily visible signs of breast diseases such as masses and calcifications. However, even with the increased ambiguity as well as difficulty level and the inclusion of a number of normal control cases, the rates of sensitivity obtained in the present study are comparable to or better than those obtained in related previous studies [38, 39] with a substantial reduction in the number of FPs per image.

8.5   REMARKS

In this chapter, the results obtained using the methods presented for the detection of architectural distortion were discussed. The effects of the use of several types of feature sets and their combinations were tested and analyzed in terms of ROC and FROC analysis. The statistical significance of the results obtained was also analyzed. Several types of cross-validation, feature selection, and pattern classification methods were used for this purpose. The results and analysis presented in this chapter demonstrate that the methods could be effectively used to assess the presence of architectural distortion in prior mammograms of interval-cancer cases.

CHAPTER 9

Concluding Remarks

Architectural distortion commonly exhibits a node-like pattern with spicules that appear to radiate from a point. However, CLS related to ligaments, ducts, vessels, and edges of parenchymal regions that overlap in the mammographic projection image could also give rise to similar patterns. As a result, the use of Gabor filters and phase-portrait analysis [23] results in the detection of not only potential sites of architectural distortion but also a number of FPs in each mammographic image. The derivation of additional features related to the textural patterns and characteristics of the regions of architectural distortion to assist in the detection of architectural distortion with high sensitivity and low FPR is the main original contribution of the present work.

Haralick's texture measures [91, 92] are widely used for texture analysis; they are based upon statistical properties of the pixels in the given image, and some of the measures may not be directly related to the intersecting structures, spiculations, and node-like patterns of architectural distortion. Some of Haralick's features are difficult to relate to the structural characteristics of an image. In this context, Laws' texture energy measures can address the limitations of Haralick's texture measures by considering the structural characteristics of a given image or ROI. In addition, if they are used together, they could form a powerful tool to perform characterization and detection of random texture and oriented patterns in a given image. The geometric transformation of the ROI being processed so as to convert spiculating patterns to waves or ripples, as presented in this book, is a prerequisite to the derivation of Laws' texture energy measures for the characterization of architectural distortion, and is an important and novel contribution.

Although several studies have been reported on fractal analysis using the Fourier power spectrum, the angular spread or distribution of power is often overlooked; only the Fourier spectral power as a function of radial frequency averaged over all angles is considered, and the information related to the angular spread of power in the Fourier domain is usually ignored. In the context of the present work, the angular spread of power in the Fourier spectrum is used to generate features for the characterization of spiculated patterns related to architectural distortion. The method presented in this book to characterize the angular spread of power is novel and effective for the analysis of oriented texture. In addition, original and novel methods were presented for the characterization and analysis of architectural distortion in prior mammograms using higher order Tsallis and Rényi entropy for quantification of the angular distribution of the magnitude and angle responses of Gabor filters, coherence, orientation strength, and power in the frequency domain.

The commonly used public databases of mammograms, the DDSM [80] and the MIAS database [14], do not have prior mammograms. As a result, the studies on architectural distortion found in the literature are specific to private and restricted datasets; the robustness of such methods has not been studied via cross-validation. An original aspect of the present work is detailed statistical analysis of the results with several feature selection and pattern classification methods; in addition, the robustness of the presented methods was studied via cross-validation among different datasets of prior, detection, as well as normal mammograms.

The automatic detection of architectural distortion in prior mammograms of interval-cancer cases, which is a difficult problem, has not been previously studied. The development of CAD techniques for the detection and localization of architectural distortion in prior mammograms was presented, which could lead to efficient detection of subtle signs of early stages of breast cancer at pre-mass-formation stages and substantial improvement in the prognosis. Accurate detection of malignancies at early stages could not only reduce health-care costs but also lessen patient suffering and increase the probability of survival from breast cancer.

Limitations exist in the present work in terms of the types or patterns of architectural distortion detected by the constrained models used. The approach of simultaneous analysis of current and prior mammograms, as recommended by radiologists for use in the screening and diagnosis of breast cancer, could be used to enhance the performance of CAD systems. The results obtained in the present work with prior mammograms of interval-cancer cases including a number of normal control cases are important and encouraging, and indicate that the presented methods have significant potential to achieve early detection of subtle signs of breast cancer in mammograms, specifically, architectural distortion.

Contrary to the state of the art in the detection of masses and calcifications, the results obtained in the present work indicate that the methods presented are not yet ready for clinical use at the moment; however, the results are better than those reported in studies on architectural distortion with commercially available CAD systems [36, 81, 83]. Furthermore, the present study is on automatic detection of architectural distortion in prior mammograms of interval-cancer cases where the corresponding regions were missed by the radiologists at the time of their original screening and interpretation: these aspects of the dataset and the problem indicate the associated high levels of complexity and difficulty.

Finally, the methods presented in this book, including the original contributions on geometric transformation, characterization of angular distribution, and use of combinations of statistical and structural properties of texture, produced statistically significant improvement in the results of ROC analysis. For the dataset of interval-cancer cases used in the present study, the methods described are capable of detecting early signs of breast cancer 15 months ahead of the time of clinical detection, on the average, with a sensitivity of 0.8 at 5.7 FP/image. The novel and original methods developed should find useful applications not only in areas of medical image processing and detection but also other areas of science and engineering.

Future lines of investigation related to the developments in the present work include:

• Methods for the reduction of false positives to less than two per image, at a sensitivity greater than 90% for the detection of architectural distortion in prior mammograms of interval-cancer cases.

- Development of more accurate methods for the identification of CLS in the mammographic image.

- Evaluation of the presented methods with larger databases.

- Modified phase-portrait models using constrained nonlinear models [209, 210] to detect patterns of architectural distortion that are not limited to node-like spiculated patterns as assumed in the previous and present studies.

- Multifractal analysis, unifying the three essential classes of FD based on morphology, entropy, and transforms using the generalized-entropy-based Rényi FD spectrum [199].

- Further reduction in the FP rate by the detection and removal of the pectoral muscle [211] in MLO views and additional conditions at the edges of the fibroglandular disk [132].

- Use of nonlinear Lyapunov exponents [212, 213] and chaos [186, 214] to model and predict the development of architectural distortion in time series of mammograms in a screening program.

- Analysis of divergence of tissue patterns in relation to expected local orientation modeled using the breast boundary for reference [43].

The development of efficient methods for the detection and localization of architectural distortion in prior mammograms is expected to lead to the detection of subtle signs of early stages of breast cancer at pre-mass-formation stages, and thereby help to improve the prognosis of breast-cancer patients.

APPENDIX A

List of Empirically Selected Parameters

Gabor filter width, τ	4 pixels
Gabor filter elongation factor, l	8
Number of Gabor filter kernels, K	180
Range of Gabor filter angle, θ	$[-\pi/2, \pi/2]$
Gaussian filter for segmentation of the breast portion	size 13×13 pixels, standard deviation 2 pixels
Morphological opening filter for segmentation of the breast portion	disk-shaped structuring element of radius 25 pixels
Gaussian filter for filtering and downsampling of the orientation field	$\sigma_f = 7$ pixels
Condition number for phase-portrait analysis	≤ 3.0
Gaussian filter for accumulation of votes in the node map	size 35×35 pixels, standard deviation 6 pixels
Window size for analysis of phase portraits	10×10 pixels
Size of the detected ROIs	128×128 pixels at 200 μm/pixel

Size of the Fourier spectrum of ROIs	256×256 pixels
The range of f used to fit the linear model in the computation of FD	[6, 96] pixels or [0.117, 1.875] mm^{-1}
Number of bins for computation of rose diagrams for the characterization of angular spread in the frequency domain	90, equally spaced over [0°, 179°]
Resolution for the computation of Haralick's features	8 bpp at 200 μm/pixel
Computation of GCM	distance, $d = 1$ pixel and angles, $\theta = 0°, 45°, 90°,$ and 135°
Geometric transformation for the application of Laws' texture energy measures	r: 2 to $\lceil \frac{(M, N)}{2} \rceil - 1$ (one-half of the smaller of the two dimensions rounded up to the smallest following integer minus one), for an ROI of size $M \times N$; θ: [0°, 359°]
Window size for computation of coherence	$P \times P, P = 15$
Number of bins for the computation of rose diagrams for the characterization of angular spread	60, equally spaced over [−90°, 89°]
Order of Tsallis entropy, q	2
Order of Rényi entropy, q	8
SHL-ANN parameters	hidden layer: tangent sigmoid activation function; output layer: pure linear activation function; training algorithm: Levenberg-Marquardt
RBF parameters	number of RBF neurons: up to 50; spread factor: range [0.1, 10]

References

[1] Centers for Disease Control and Prevention (CDC). Cancer among women, May 2011. Available at http://www.cdc.gov/cancer/dcpc/data/women.htm, last accessed March 21, 2012. Cited on page(s) 1

[2] J. Tang, R. M. Rangayyan, J. Xu, I. E. Naqa, and Y. Yang. Computer-aided detection and diagnosis of breast cancer with mammography: Recent advances. *IEEE Transactions on Information Technology in Biomedicine*, 13(2):236–251, March 2009. DOI: 10.1109/TITB.2008.2009441 Cited on page(s) 1, 10, 12, 13, 18

[3] Canadian Cancer Society. Breast cancer statistics, May 2010. Available at `http://www.cancer.ca/Canada-wide/Aboutcancer/Cancerstatistics/Statsataglance/Breastcancer.aspx?sc_lang=en`, accessed on 17 December, 2010. Cited on page(s) 1

[4] M. A. Schneider. Better detection: Improving our chances. In M. J. Yaffe, editor, *Digital Mammography: 5th International Workshop on Digital Mammography*, pages 3–6, Toronto, ON, Canada, June 2001. Medical Physics Publishing. Cited on page(s) 1

[5] A. Jemal, L. X. Clegg, E. Ward, L. A. G. Ries, X. Wu, P. M. Jamison, P. A. Wingo, H. L. Howe, R. N. Anderson, and B. K. Edwards. Annual report to the nation on the status of cancer, 1975-2001, with a special feature regarding survival. *Cancer*, 101(1):3–27, 2004. DOI: 10.1002/cncr.20288 Cited on page(s) 1

[6] C. Di Maggio. State of the art of current modalities for the diagnosis of breast lesions. *European Journal on Nuclear Medicine and Molecular Imaging*, 31, supplement 1:S56—S69, June 2004. DOI: 10.1007/978-3-540-36781-9_9 Cited on page(s) 1

[7] A. K. Hackshaw and E. A. Paul. Breast self-examination and death from breast cancer: a meta-analysis. *British Journal of Cancer*, 88:1047–1053, 2003. DOI: 10.1038/sj.bjc.6600847 Cited on page(s) 1

[8] M. J. Homer. *Mammographic Interpretation: A Practical Approach*. McGraw-Hill, New York, NY, 2nd edition, 1997. Cited on page(s) 1, 3, 5, 15, 16, 87

[9] S. Baeg and N. Kehtarnavaz. Classification of breast mass abnormalities using denseness and architectural distortion. *Electronic Letters on Computer Vision and Image Analysis*, 1(1):1–20, August 2002. Cited on page(s) 1

[10] B. Cady and M. Chung. Mammographic screening: No longer controversial. *American Journal of Clinical Oncology*, 28(1):1–4, February 2005. DOI: 10.1097/01.coc.0000150720.15450.05 Cited on page(s) 2

[11] G. Cardenosa. *Breast Imaging Companion*. Lippincot-Raven, Philadelphia, PA, 1st edition, 1997. Cited on page(s) 2, 4

[12] R. M. Rangayyan. *Biomedical Image Analysis*. CRC Press, Boca Raton, FL, 2005. Cited on page(s) 3, 5, 15, 27, 28, 69, 71, 72, 82, 87, 90, 92, 93, 97, 98, 99, 102, 113

[13] American College of Radiology (ACR). *Illustrated Breast Imaging Reporting and Data System (BI-RADS®)*. American College of Radiology, Reston, VA, 3rd edition, 1998. Cited on page(s) 3, 4, 15, 46, 54, 87

[14] J. Suckling, J. Parker, D. R. Dance, S. Astley, I. Hutt, C. R. M. Boggis, I. Ricketts, E. Stamakis, N. Cerneaz, S.-L. Kok, P. Taylor, D. Betal, and J. Savage. The Mammographic Image Analysis Society Digital Mammogram Database. In A. G. Gale, S. M. Astley, D. D. Dance, and A. Y. Cairns, editors, *Digital Mammography: Proceedings of the 2nd International Workshop on Digital Mammography*, pages 375–378, York, UK, July 1994. Elsevier. Cited on page(s) 3, 4, 5, 6, 7, 16, 17, 19, 21, 136, 139

[15] A. M. Knutzen and J. J. Gisvold. Likelihood of malignant disease for various categories of mammographically detected, nonpalpable breast lesions. *Mayo Clinic Proceedings*, 68:454–460, 1993. DOI: 10.1016/S0025-6196(12)60194-3 Cited on page(s) 6, 15

[16] H. C. Burrell, D. M. Sibbering, A. R. M. Wilson, S. E. Pinder, A. J. Evans, L. J. Yeoman, C. W. Elston, I. O. Ellis, R. W. Blamey, and J. F. R. Robertson. Screening interval breast cancers: Mammographic features and prognostic factors. *Radiology*, 199(4):811–817, 1996. Cited on page(s) 6, 16

[17] C. E. Metz. ROC methodology in radiologic imaging. *Investigative Radiology*, 21:720–733, 1986. DOI: 10.1097/00004424-198609000-00009 Cited on page(s) 9, 63, 66

[18] C. E. Metz. Basic principles of ROC analysis. *Seminars in Nuclear Medicine*, VIII(4):283–298, 1978. DOI: 10.1016/S0001-2998(78)80014-2 Cited on page(s) 9, 63, 66

[19] H. Bornefalk and A. B. Hermansson. On the comparison of FROC curves in mammography CAD systems. *Medical Physics*, 32(2):412–417, January 2005. DOI: 10.1118/1.1844433 Cited on page(s) 10

[20] H. Miller. The FROC curve: A representation of the observer's performance for the method of free response. *Journal of the Acoustical Society of America*, 46(6B):1473–1476, December 1969. DOI: 10.1121/1.1911889 Cited on page(s)

[21] D. P. Chakraborty. Statistical power in observer-performance studies: Comparison of the receiver operating characteristic and free-response methods in tasks involving localization. *Academic Radiology*, 9(2):147–156, February 2002. DOI: 10.1016/S1076-6332(03)80164-2 Cited on page(s) 10, 67, 125

[22] R. E. Bird, T. W. Wallace, and B. C. Yankaskas. Analysis of cancers missed at screening mammography. *Radiology*, 184(3):613–617, 1992. Cited on page(s) 10, 12, 18

[23] F. J. Ayres, R. M. Rangayyan, and J. E. L. Desautels. *Analysis of Oriented Texture: with Applications to the Detection of Architectural Distortion in Mammograms*. Morgan & Claypool Publishers, 2011. DOI: 10.2200/S00301ED1V01Y201010BME038 Cited on page(s) 10, 11, 13, 27, 32, 35, 36, 38, 39, 40, 41, 42, 45, 47, 139

[24] J. A. A. M. van Dijck, A. L. M. Verbeek, J. H. C. L. Hendriks, and R. Holland. The current detectability of breast cancer in a mammographic screening program. *Cancer*, 72(6):1933–1938, 1993. Cited on page(s) 10, 12, 18

[25] R. G. Blanks, M. G. Wallis, and S. M. Moss. A comparison of cancer detection rates achieved by breast cancer screening programmes by number of readers, for one and two view mammography: Results from the UK National Health Service Breast Screening Programme. *Journal of Medical Screening*, 5(4):195–201, 1998. DOI: 10.1136/jms.5.4.195 Cited on page(s) 10

[26] K. Doi. Diagnostic imaging over the last 50 years: research and development in medical imaging science and technology. *Physics in Medicine and Biology*, 51:R5—R27, June 2006. DOI: 10.1088/0031-9155/51/13/R02 Cited on page(s) 10, 12, 18

[27] K. Doi. Computer-aided diagnosis in medical imaging: historical review, current status and future potential. *Computerized Medical Imaging and Graphics*, 31:198–211, 2007. DOI: 10.1016/j.compmedimag.2007.02.002 Cited on page(s) 10, 11

[28] R. M. Rangayyan, F. J. Ayres, and J. E. L. Desautels. A review of computer-aided diagnosis of breast cancer: Toward the detection of subtle signs. *Journal of the Franklin Institute*, 344:312–348, 2007. DOI: 10.1016/j.jfranklin.2006.09.003 Cited on page(s) 10, 12, 13, 18

[29] M. Nemoto, S. Honmura, A. Shimizu, D. Furukawa, H. Kobatake, and S. Nawano. A pilot study of architectural distortion detection in mammograms based on characteristics of line shadows. *International Journal of Computer Assisted Radiology and Surgery*, 4(1):27–36, January 2009. DOI: 10.1007/s11548-008-0267-9 Cited on page(s) 10, 17, 21, 136

[30] R2 Technology website, http://www.hologic.com/en/breast-imaging, accessed on January 20, 2011. Cited on page(s) 11

[31] iCAD website, http://www.icadmed.com/, accessed on January 21, 2011. Cited on page(s) 11

[32] S. M. Astley and F. J. Gilbert. Computer-aided detection in mammography. *Clinical Radiology*, 59:390–399, 2004. DOI: 10.1016/j.crad.2003.11.017 Cited on page(s) 11

[33] S. Ciatto, M. R. Del Turco, G. Risso, S. Catarzi, R. Bonaldi, V. Viterbo, P. Gnutti, B. Guglielmoni, L. Pinelli, A. Pandiscia, F. Navarra, A. Lauria, R. Palmiero, and P. L. Indovina. Comparison of standard reading and computer aided detection (CAD) on a national proficiency test of screening mammography. *European Journal of Radiology*, 45:135–138, 2003. DOI: 10.1016/S0720-048X(02)00011-6 Cited on page(s) 11

[34] T. W. Freer and M. J. Ulissey. Screening mammography with computer-aided detection: Prospective study of 12,860 patients in a community breast center. *Radiology*, 220:781–786, 2001. DOI: 10.1148/radiol.2203001282 Cited on page(s) 11

[35] J. J. Fenton, S. H. Taplin, P. A. Carney, L. Abraham, E. A. Sickles, C. D'Orsi, E. A. Berns, G. Cutter, R. E. Hendrick, W. E. Barlow, and J. G. Elmore. Influence of computer-aided detection on performance of screening mammography. *The New England Journal of Medicine*, 356(14):1399–1409, April 2007. DOI: 10.1056/NEJMoa066099 Cited on page(s) 11

[36] J. A. Baker, E. L. Rosen, J. Y. Lo, E. I. Gimenez, R. Walsh, and M. S. Soo. Computer-aided detection (CAD) in screening mammography: Sensitivity of commercial CAD systems for detecting architectural distortion. *American Journal of Roentgenology*, 181:1083–1088, 2003. Cited on page(s) 12, 15, 18, 140

[37] M. J. M. Broeders, N. C. Onland-Moret, H. J. T. M. Rijken, J. H. C. L. Hendriks, A. L. M. Verbeek, and R. Holland. Use of previous screening mammograms to identify features indicating cases that would have a possible gain in prognosis following earlier detection. *European Journal of Cancer*, 39:1770–1775, 2003. DOI: 10.1016/S0959-8049(03)00311-3 Cited on page(s) 12, 15

[38] R. M. Rangayyan, S. Prajna, F. J. Ayres, and J. E. L. Desautels. Detection of architectural distortion in mammograms acquired prior to the detection of breast cancer using Gabor filters, phase portraits, fractal dimension, and texture analysis. *International Journal of Computer Assisted Radiology and Surgery*, 2(6):347–361, April 2008. DOI: 10.1007/s11548-007-0143-z Cited on page(s) 12, 13, 18, 22, 23, 24, 36, 39, 92, 93, 136, 138

[39] R. M. Rangayyan, S. Banik, and J. E. L. Desautels. Computer-aided detection of architectural distortion in prior mammograms of interval cancer. *Journal of Digital Imaging*, 23(5):611–631, October 2010. DOI: 10.1007/s10278-009-9257-x Cited on page(s) 22, 25, 39, 113, 120, 135, 136, 138

[40] S. Banik, R. M. Rangayyan, and J. E. L. Desautels. Detection of architectural distortion in prior mammograms. *IEEE Transactions on Medical Imaging*, 30(2):279–294, February 2011. DOI: 10.1109/TMI.2010.2076828 Cited on page(s) 12, 18, 22, 25, 113, 115, 121, 135

[41] M. Sameti, R. K. Ward, J. Morgan-Parkes, and B. Palcic. Image feature extraction in the last screening mammograms prior to detection of breast cancer. *IEEE Journal of Selected Topics in Signal Processing*, 3(1):46–52, February 2009. DOI: 10.1109/JSTSP.2008.2011163 Cited on page(s) 18, 22, 24, 136

[42] S. Banik, R. M. Rangayyan, and J. E. L. Desautels. Measures of angular spread and entropy for the detection of architectural distortion in prior mammograms. *International Journal of Computer Assisted Radiology and Surgery*, vol. 8(1), pp. 121–134, January 2013. DOI: 10.1007/s11548-012-0681-x Cited on page(s) 12, 113, 114, 135

[43] R. M. Rangayyan, S. Banik, J. Chakraborty, S. Mukhopadhyay, and J. E. L. Desautels. Measures of divergence of oriented patterns for the detection of architectural distortion in prior mammograms. *International Journal of Computer Assisted Radiology and Surgery*, 2012. In press. DOI: 10.1007/s11548-012-0793-3 Cited on page(s) 12, 18, 135, 141

[44] J. H. Sumkin, B. L. Holbert, J. S. Herrmann, C. A. Hakim, M. A. Ganott, W. R. Poller, R. Shah, L. A. Hardesty, and D. Gur. Optimal reference mammography: A comparison of mammograms obtained 1 and 2 years before the present examination. *American Journal of Roentgenology*, 180:343–346, 2003. Cited on page(s) 12, 13, 23

[45] C. Varela, N. Karssemeijer, J. H. C. L. Hendriks, and R. Holland. Use of prior mammograms in the classification of benign and malignant masses. *European Journal of Radiology*, 56:248–255, 2005. DOI: 10.1016/j.ejrad.2005.04.007 Cited on page(s) 23

[46] A. S. Majid, E. S. de Paredes, R. D. Doherty, N. R. Sharma, and X. Salvador. Missed breast carcinoma: Pitfalls and pearls. *RadioGraphics*, 23:881–895, 2003. DOI: 10.1148/rg.234025083 Cited on page(s) 12, 13, 23

[47] H. Alto, R. M. Rangayyan, and J. E. L. Desautels. Content-based retrieval and analysis of mammographic masses. *Journal of Electronic Imaging*, 14(2):Article number 023016, pp. 1–17, 2005. DOI: 10.1117/1.1902996 Cited on page(s) 13

[48] S. K. Kinoshita, P. M. de Azevedo-Marques, R. R. Pereira, J. A. H. Rodrigues, and R. M. Rangayyan. Content-based retrieval of mammograms using visual features related to breast density patterns. *Journal of Digital Imaging*, 20(2):172–190, June 2007. DOI: 10.1007/s10278-007-9004-0 Cited on page(s)

[49] P. M. de Azevedo-Marques, N. A. Rosa, A. J. M. Traina, C. Traina Jr., S. K. Kinoshita, and R. M. Rangayyan. Reducing the semantic gap in content-based image retrieval in mammography with relevance feedback and inclusion of expert knowledge. *International Journal of Computer Assisted Radiology and Surgery*, 3(1–2):123–130, June 2008. DOI: 10.1007/s11548-008-0154-4 Cited on page(s) 13

[50] D. Guliato, E. V. de Melo, R. S. Bôaventura, and R. M. Rangayyan. AMDI - indexed atlas of digital mammograms that integrates case studies, e-learning, and research systems via the web. In J. S. Suri and R. M. Rangayyan, editors, *Recent Advances in Breast Imaging, Mammography, and Computer-aided Diagnosis of Breast Cancer*, pages 529–555. SPIE Press, Bellingham, WA, 2006. DOI: 10.1117/3.651880 Cited on page(s) 13

[51] D. Guliato, R. S. Bôaventura, M. Maia, R. M. Rangayyan, M. S. Simedo, and T. A. A. Macedo. INDIAM—an e-learning system for the interpretation of mammograms. *Journal of Digital Imaging*, 22(4):405–420, August 2009. DOI: 10.1007/s10278-008-9111-6 Cited on page(s) 13

[52] M. R. Smith, X. Wang, and R. M. Rangayyan. Evaluation of the sensitivity of a medical data-mining application to the number of elements in small databases. *Biomedical Signal Processing and Control*, 4(3):262–268, 2009. DOI: 10.1016/j.bspc.2009.04.001 Cited on page(s) 13

[53] E. A. Sickles. Mammographic features of 300 consecutive nonpalpable breast cancers. *American Journal of Roentgenology*, 146(4):661–663, 1986. Cited on page(s) 14, 15

[54] T. Matsubara, T. Ichikawa, T. Hara, H. Fujita, S. Kasai, T. Endo, and T. Iwase. Novel method for detecting mammographic architectural distortion based on concentration of mammary gland. *International Congress Series, Elsevier B.V.*, 1268:867–871, June 2004. Proceedings of the 18th International Congress and Exhibition on Computer Assisted Radiology and Surgery (CARS2004). DOI: 10.1016/j.ics.2004.03.103 Cited on page(s) 15, 16, 136

[55] G. D. Tourassi, D. M. Delong, and C. E. Floyd Jr. A study on the computerized fractal analysis of architectural distortion in screening mammograms. *Physics in Medicine and Biology*, 51(5):1299–1312, 2006. DOI: 10.1088/0031-9155/51/5/018 Cited on page(s) 15, 17, 20, 90, 93, 94, 136

[56] B. C. Yankaskas, M. J. Schell, R. E. Bird, and D. A. Desrochers. Reassessment of breast cancers missed during routine screening mammography: a community based study. *American Journal of Roentgenology*, 177:535–541, 2001. Cited on page(s) 16

[57] H. Burrell, A. Evans, A. Wilson, and S. Pinder. False-negative breast screening assessment: what lessons we can learn? *Clinical Radiology*, 56:385–388, 2001. DOI: 10.1053/crad.2001.0662 Cited on page(s) 16

[58] N. Karssemeijer and G. M. te Brake. Detection of stellate distortions in mammograms. *IEEE Transactions on Medical Imaging*, 15(5):611–619, October 1996. DOI: 10.1109/42.538938 Cited on page(s) 16, 19, 41, 42, 136

[59] N. R. Mudigonda and R. M. Rangayyan. Texture flow-field analysis for the detection of architectural distortion in mammograms. In A. G. Ramakrishnan, editor, *Proceedings of Biovision*, pages 76–81, Bangalore, India, December 2001. Cited on page(s) 16

[60] F. J. Ayres and R. M. Rangayyan. Reduction of false positives in the detection of architectural distortion in mammograms by using a geometrically constrained phase portrait model. *International Journal of Computer Assisted Radiology and Surgery*, 1:361–369, 2007. DOI: 10.1007/s11548-007-0072-x Cited on page(s) 16, 21, 32, 36, 39, 41, 45, 46

[61] F. J. Ayres and R. M. Rangayyan. Characterization of architectural distortion in mammograms. In *Proceedings of the 25th Annual International Conference of the IEEE Engineering in Medicine and Biology Society (CD-ROM)*, pages 886–889, Cancún, Mexico, September 2003. DOI: 10.1109/MEMB.2005.1384102 Cited on page(s)

[62] F. J. Ayres and R. M. Rangayyan. Detection of architectural distortion in mammograms via analysis of phase portraits and curvilinear structures. In J. Hozman and P. Kneppo, editors, *Proceedings of EMBEC'05: 3rd European Medical & Biological Engineering Conference*, volume 11, pages 1768–1773, Prague, Czech Republic, November 2005. Cited on page(s) 41

[63] F. J. Ayres and R. M. Rangayyan. Characterization of architectural distortion in mammograms. *IEEE Engineering in Medicine and Biology Magazine*, 24(1):59–67, January 2005. DOI: 10.1109/MEMB.2005.1384102 Cited on page(s) 16, 21, 22, 36

[64] R. M. Rangayyan and F. J. Ayres. Gabor filters and phase portraits for the detection of architectural distortion in mammograms. *Medical and Biological Engineering and Computing*, 44:883–894, August 2006. DOI: 10.1007/s11517-006-0088-3 Cited on page(s) 16, 22, 27, 28, 31, 32, 36, 39, 41, 45

[65] T. Matsubara, T. Ichikawa, T. Hara, H. Fujita, S. Kasai, T. Endo, and T. Iwase. Automated detection methods for architectural distortions around skinline and within mammary gland on mammograms. In H. U. Lemke, M. W. Vannier, K. Inamura, A. G. Farman, K. Doi, and J. H. C. Reiber, editors, *International Congress Series: Proceedings of the 17th International Congress and Exhibition on Computer Assisted Radiology and Surgery*, pages 950–955, London, UK, June 2003. Elsevier. Cited on page(s) 16, 20

[66] T. Ichikawa, T. Matsubara, T. Hara, H. Fujita, T. Endo, and T. Iwase. Automated detection method for architectural distortion areas on mammograms based on morphological processing and surface analysis. In J. M. Fitzpatrick and M. Sonka, editors, *Proceedings of SPIE Medical Imaging 2004: Image Processing*, pages 920–925, San Diego, CA, February 2004. SPIE. Cited on page(s) 16, 20, 41

[67] T. Hara, T. Makita, T. Matsubara, H. Fujita, Y. Inenaga, T. Endo, and T. Iwase. Automated detection method for architectural distortion with spiculation based on distribution assessment of mammary gland on mammogram. In S. M. Astley, M. Brady, C. Rose, and R. Zwiggelaar, editors, *Digital Mammography / IWDM*, volume 4046 of *Lecture Notes in Computer Science*, pages 370–375, Manchester, UK, June 2006. Cited on page(s) 16, 20

[68] T. Matsubara, T. Hara, H. Fujita, T. Endo, and T. Iwase. Automated detection method for mammographic spiculated architectural distortion based on surface analysis. In *Proceedings of the 22nd International Congress and Exhibition on Computer Assisted Radiology and Surgery (CARS2008)*, volume 3(1), pages S176—S177, Barcelona, Spain, June 2008. DOI: 10.1117/12.535116 Cited on page(s) 16, 20

[69] Q. Guo, J. Shao, and V. Ruiz. Investigation of support vector machine for the detection of architectural distortion in mammographic images. *Journal of Physics: Conference Series*, 15:88–94, 2005. DOI: 10.1088/1742-6596/15/1/015 Cited on page(s) 16, 19

[70] Q. Guo, J. Shao, and V. F. Ruiz. Characterization and classification of tumor lesions using computerized fractal-based texture analysis and support vector machines in digital mammograms. *International Journal of Computer Assisted Radiology and Surgery*, 4(1):11–25, January 2009. DOI: 10.1007/s11548-008-0276-8 Cited on page(s) 17, 19, 93, 136

[71] M. P. Sampat and A. C. Bovik. Detection of spiculated lesions in mammograms. In *Proceedings of the 25th Annual International Conference of the IEEE Engineering in Medicine and Biology Society (CD-ROM)*, pages 810–813, Cancún, Mexico, September 2003. DOI: 10.1109/IEMBS.2003.1279888 Cited on page(s) 17

[72] M. P. Sampat, G. J. Whitman, M. K. Markey, and A. C. Bovik. Evidence based detection of spiculated masses and architectural distortion. In J. M. Fitzpatrick and J. M. Reinhardt, editors, *Proceedings of SPIE Medical Imaging 2005: Image Processing*, volume 5747, pages 26–37, San Diego, CA, April 2005. Cited on page(s) 17, 19

[73] M. P. Sampat, M. K. Markey, and A. C. Bovik. Measurement and detection of spiculated lesions. In *IEEE Southwest Symposium on Image Analysis and Interpretation*, pages 105–109. IEEE Computer Society, March 2006. DOI: 10.1109/SSIAI.2006.1633731 Cited on page(s) 17, 19, 136

[74] S. Özekes, O. Osman, and A. Y. Çamurcu. Computerized detection of architectural distortions in digital mammograms. In *Proceedings of the 19th International Conference on Computer Assisted Radiology and Surgery (CARS2005)*, volume 1281, page 1396, Berlin, Germany, 2005. Cited on page(s) 17, 21

[75] M. Aguilar, E. Anguiano, and M. A. Pancorbo. Fractal characterization by frequency analysis: II. A new method. *Journal of Microscopy*, 172:233–238, 1993.
DOI: 10.1111/j.1365-2818.1993.tb03417.x Cited on page(s) 17, 22, 92, 93, 94

[76] E. Anguiano, M. A. Pancorbo, and M. Aguilar. Fractal characterization by frequency analysis: I. Surfaces. *Journal of Microscopy*, 172:223–232, 1993.
DOI: 10.1111/j.1365-2818.1993.tb03416.x Cited on page(s) 17, 22, 92, 93, 94

[77] N. Eltonsy, G. Tourassi, and A. Elmaghraby. Investigating performance of a morphology-based CAD scheme in detecting architectural distortion in screening mammograms. In H. U. Lemke, K. Inamura, K. Doi, M. W. Vannier, and A. G. Farman, editors, *Proceedings of the 20th International Congress and Exhibition on Computer Assisted Radiology and Surgery (CARS 2006)*, pages 336–338, Osaka, Japan, June 2006. Springer. Cited on page(s) 17, 20, 136

[78] R. Nakayama, R. Watanabe, T. Kawamura, T. Takada, K. Yamamoto, and K. Takeda. Computer-aided diagnosis scheme for detection of architectural distortion on mammograms using multiresolution analysis. In *Proceedings of the 22nd International Congress and Exhibition on Computer Assisted Radiology and Surgery (CARS 2008)*, volume 3(1), pages S418–S419, Barcelona, Spain, June 2008. Cited on page(s) 17, 21, 136

[79] M. Jasionowska, A. Przelaskowski, A. Rutczynska, and A. Wroblewska. A two-step method for detection of architectural distortions in mammograms. In *Information Technologies in Biomedicine*, volume 69, pages 73–84. Springer Berlin / Heidelberg, 2010.
DOI: 10.1007/978-3-642-13105-9_8 Cited on page(s) 17, 136

[80] M. Heath, K. Bowyer, D. Kopans, R. Moore, and W. P. Kegelmeyer. The Digital Database for Screening Mammography. In M.J. Yaffe, editor, *Proceedings of the Fifth International Workshop on Digital Mammography*, pages 212–218. Medical Physics Publishing, 2001. Cited on page(s) 18, 20, 136, 139

[81] L. J. W. Burhenne, S. A. Wood, C. J. D'Orsi, S. A. Feig, D. B. Kopans, K. F. O'Shaughnessy, E. A. Sickles, L. Tabar, C. J. Vyborny, and R. A. Castellino. Potential contribution of computer-aided detection to the sensitivity of screening mammography. *Radiology*, 215(2):554–562, 2000. Cited on page(s) 18, 140

[82] W. P. Evans, L. J. W. Burhenne, L. Laurie, K. F. O'Shaughnessy, and R. A. Castellino. Invasive lobular carcinoma of the breast: Mammographic characteristics and computer-aided detection. *Radiology*, 225(1):182–189, 2002. DOI: 10.1148/radiol.2251011029 Cited on page(s) 18, 23

[83] R. L. Birdwell, D. M. Ikeda, K. F. O'Shaughnessy, and E. A. Sickles. Mammographic characteristics of 115 missed cancers later detected with screening mammography and the potential utility of computer-aided detection. *Radiology*, 219(1):192–202, 2001. Cited on page(s) 18, 140

[84] S. Banik, R. M. Rangayyan, and J. E. L. Desautels. Detection of architectural distortion in prior mammograms using fractal analysis and angular spread of power. In N. Karssemeijer and R. M. Summers, editors, *Proceedings of SPIE Medical Imaging 2010: Computer Aided Diagnosis*, pages 762408–1–9, San Diego, CA, February 2010. Cited on page(s) 18, 22, 39, 108, 121, 135

[85] R. M. Rangayyan, S. Banik, S. Prajna, and J. E. L. Desautels. Detection of architectural distortion in prior mammograms of interval-cancer cases. In *Proceedings of the 23rd International*

Congress and Exhibition: Computer Assisted Radiology and Surgery, pages S171—S173, Berlin, Germany, June 2009. DOI: 10.1109/TMI.2010.2076828 Cited on page(s)

[86] S. Banik, R. M. Rangayyan, and J. E. L. Desautels. Detection of architectural distortion in prior mammograms of interval-cancer cases with neural networks. In *Proceedings of the 31st Annual International Conference of the IEEE Engineering in Medicine and Biology Society*, pages 6667–6670, Minneapolis, MN, September 2009. DOI: 10.1117/12.843840 Cited on page(s)

[87] S. Banik, R. M. Rangayyan, and J. E. L. Desautels. Detection of architectural distortion in prior mammograms of interval cancer using Laws' texture energy measures. In *Proceedings of the 24th International Congress and Exhibition: Computer Assisted Radiology and Surgery*, volume 5(1), pages S200—S201, Geneva, Switzerland, June 2010. DOI: 10.1109/TMI.2010.2076828 Cited on page(s) 18, 22, 39, 97, 120

[88] A. R. Rao. *A Taxonomy for Texture Description and Identification*. Springer-Verlag, New York, NY, 1990. DOI: 10.1007/978-1-4613-9777-9 Cited on page(s) 22, 31, 32, 34, 36, 110, 111

[89] A. R. Rao and R. C. Jain. Computerized flow field analysis: Oriented texture fields. *IEEE Transactions on Pattern Analysis and Machine Intelligence*, 14(7):693–709, July 1992. DOI: 10.1109/34.142908 Cited on page(s) 22, 31, 32, 34, 36, 38, 110

[90] K. I. Laws. Rapid texture identification. In *Proceedings of SPIE Vol. 238: Image Processing for Missile Guidance*, pages 376–380, San Diego, CA, 1980. DOI: 10.1117/12.959169 Cited on page(s) 22, 102

[91] R. M. Haralick. Statistical and structural approaches to texture. *Proceedings of the IEEE*, 67:786–804, May 1979. DOI: 10.1109/PROC.1979.11328 Cited on page(s) 22, 98, 99, 139

[92] R. M. Haralick, K. Shanmugam, and I. Dinstein. Textural features for image classification. *IEEE Transactions on Systems, Man, Cybernetics*, 3(6):610–622, 1973. DOI: 10.1109/TSMC.1973.4309314 Cited on page(s) 22, 98, 99, 113, 139

[93] M. Sameti, J. Morgan-Parkes, R. K. Ward, and B. Palcic. Classifying image features in the last screening mammograms prior to detection of a malignant mass. In N. Karssemeijer, M. Thijssen, J. Hendriks, and L. van Erning, editors, *Proceedings of the 4th International Workshop on Digital Mammography*, pages 127–134, Nijmegen, The Netherlands, June 1998. DOI: 10.1007/978-94-011-5318-8 Cited on page(s) 22

[94] N. Petrick, H. P. Chan, B. Sahiner, M. A. Helvie, and S. Paquerault. Evaluation of an automated computer-aided diagnosis system for the detection of masses on prior mammograms. In *Proceedings of SPIE Volume 3979, Medical Imaging 2000: Image Processing*, pages 967–973, San Diego, CA, 2000. DOI: 10.1117/12.387600 Cited on page(s) 22, 25

[95] B. Zheng, W. F. Good, D. R. Armfield, C. Cohen, T. Hertzberg, J. H. Sumkin, and D. Gur. Performance change of mammographic CAD schemes optimized with most-recent and prior image databases. *Academic Radiology*, 10:283–288, 2003. DOI: 10.1016/S1076-6332(03)80102-2 Cited on page(s) 22, 24, 133

[96] E. S. Burnside, E. A. Sickles, R. E. Sohlich, and K. E. Dee. Differential value of comparison with previous examinations in diagnostic versus screening mammography. *American Journal of Roentgenology*, 179:1173–1177, 2002. Cited on page(s) 23

[97] S. Ciatto, M. R. Del Turco, P. Burke, C. Visioli, E. Paci, and M Zappa. Comparison of standard and double reading and computer-aided detection (CAD) of interval cancers at prior negative screening mammograms: blind review. *British Journal of Cancer*, 89:1645–1649, 2003. DOI: 10.1038/sj.bjc.6601356 Cited on page(s) 23

[98] K. Moberg, N. Bjurstam, B. Wilczek, L. Rostgård, E. Egge, and C. Muren. Computer assisted detection of interval breast cancers. *European Journal of Radiology*, 39:104–110, 2001. DOI: 10.1016/S0720-048X(01)00291-1 Cited on page(s) 23

[99] D. M. Ikeda, R. L. Birdwell, K. F. O'Shaughnessy, E. A. Sickles, and R. J. Brenner. Computer-aided detection output on 172 subtle findings on normal mammograms previously obtained in women with breast cancer detected at follow-up screening mammography. *Radiology*, 230:811–819, 2004. DOI: 10.1148/radiol.2303030254 Cited on page(s) 23

[100] L. Garvican and S. Field. A pilot evaluation of the R2 Image Checker System and users' response in the detection of interval breast cancers on previous screening films. *Clinical Radiology*, 56:833–837, 2001. DOI: 10.1053/crad.2001.0776 Cited on page(s) 23

[101] F. J. Ayres and R. M. Rangayyan. Design and performance analysis of oriented feature detectors. *Journal of Electronic Imaging*, 16(2):12 pages, April 2007. article number 023007. DOI: 10.1117/1.2728751 Cited on page(s) 27, 28, 31

[102] S. Chaudhuri, H. Nguyen, R. M. Rangayyan, S. Walsh, and C. B. Frank. A Fourier domain directional filtering method for analysis of collagen alignment in ligaments. *IEEE Transactions on Biomedical Engineering*, 34(7):509–518, 1987. DOI: 10.1109/TBME.1987.325980 Cited on page(s) 27

[103] P. Embree and J. P. Burg. Wide-band velocity filtering—the pie slice process. *Geophysics*, 28:948–974, 1963. DOI: 10.1190/1.1439310 Cited on page(s)

[104] S. Treitel, J. L. Shanks, and C. W. Frasier. Some aspects of fan filtering. *Geophysics*, 32:789–806, 1967. DOI: 10.1190/1.1439889 Cited on page(s)

[105] V. Bezvoda, J. Ježek, and K. Segeth. FREDPACK- A program package for linear filtering in the frequency domain. *Computers & Geosciences*, 16(8):1123–1154, 1990. DOI: 10.1016/0098-3004(90)90053-V Cited on page(s) 27

[106] W. T. Freeman and E. H. Adelson. The design and use of steerable filters. *IEEE Transactions on Pattern Analysis and Machine Intelligence*, 13(9):891–906, September 1991. DOI: 10.1109/34.93808 Cited on page(s) 27, 41

[107] D. Gabor. Theory of communication. *Journal of the Institute of Electrical Engineers*, 93:429–457, 1946. Cited on page(s) 27

[108] C. K. Chui. *An Introduction to Wavelets*, volume 1 of *Wavelet Analysis and its Applications*. Academic Press, San Diego, CA, 1992. Cited on page(s)

[109] B. S. Manjunath and W. Y. Ma. Texture features for browsing and retrieval of image data. *IEEE Transactions on Pattern Analysis and Machine Intelligence*, 18(8):837–842, 1996. DOI: 10.1109/34.531803 Cited on page(s) 27, 28

[110] R. J. Ferrari, R. M. Rangayyan, J. E. L. Desautels, and A. F. Frère. Analysis of asymmetry in mammograms via directional filtering with Gabor wavelets. *IEEE Transactions on Medical Imaging*, 20(9):953–964, 2001. DOI: 10.1109/42.952732 Cited on page(s) 27, 28, 41

[111] L. T. Bruton and N. R. Bartley. Using nonessential singularities of the second kind in two-dimensional filter design. *IEEE Transactions on Circuits and Systems*, 36(1):113–116, 1989. DOI: 10.1109/31.16572 Cited on page(s) 27

[112] N. Merlet and J. Zerubia. New prospects in line detection by dynamic programming. *IEEE Transactions on Pattern Analysis and Machine Intelligence*, 18(4):426–431, April 1996. DOI: 10.1109/34.491623 Cited on page(s) 27

[113] D. S. Guru, B. H. Shekar, and P. Nagabhushan. A simple and robust line detection algorithm based on small eigenvalue analysis. *Pattern Recognition Letters*, 25:1–13, 2003. DOI: 10.1016/j.patrec.2003.08.007 Cited on page(s) 27

[114] C. R. Wylie and L. C. Barrett. *Advanced Engineering Mathematics*. McGraw-Hill, New York, NY, 6th edition, 1995. Cited on page(s) 31, 32

[115] S. Kirkpatrick, C. D. Gelatt, and M. P. Vecchi. Optimization by simulated annealing. *Science*, 220(4598):671–680, 1983. DOI: 10.1126/science.220.4598.671 Cited on page(s) 35, 45

[116] F. J. Ayres and R. M. Rangayyan. An iterative linear algorithm for the analysis of oriented patterns. In E. R. Dougherty, J. T. Astola, and K. O. Egiazarian, editors, *Electronic Imaging 2004*, volume 5298, pages 232–241, San Jose, CA, 2004. SPIE. Cited on page(s) 35

[117] F. J. Ayres and R. M. Rangayyan. Optimization procedures for the estimation of phase portrait parameters of orientation fields. In E. R. Dougherty, J. T. Astola, K. O. Egiazarian, N. M. Nasrabadi, and S. A. Rizvi, editors, *Electronic Imaging 2006*, volume 6064, San Jose, CA, 2006. SPIE. Cited on page(s) 35

[118] N. Gershenfeld. *The Nature of Mathematical Modeling.* Cambridge University Press, Cambridge, UK, 1999. Cited on page(s) 35, 45

[119] D. W. Marquardt. An algorithm for the least-squares estimation of nonlinear parameters. *Journal of the Society for Industrial and Applied Mathematics*, 11:431–441, 1963. DOI: 10.1137/0111030 Cited on page(s) 35

[120] M. Galassi, J. Davies, J. Theiler, B. Gough, G. Jungman, M. Booth, and F. Rossi. *GNU Scientific Library: Reference Manual.* Network Theory, Bristol, UK, 2nd edition, 2003. Cited on page(s) 35

[121] N. Otsu. A threshold selection method from gray-level histograms. *IEEE Transactions on Systems, Man, and Cybernetics*, 9(1):62–66, 1979. DOI: 10.1109/TSMC.1979.4310076 Cited on page(s) 39

[122] R.C. Gonzalez and R.E. Woods. *Digital Image Processing.* Prentice-Hall, Upper Saddle River, NJ, 2nd edition, 2002. Cited on page(s) 39

[123] R. J. Ferrari, R. M. Rangayyan, J. E. L. Desautels, R. A. Borges, and A. F. Frère. Identification of the breast boundary in mammograms using active contour models. *Medical and Biological Engineering and Computing*, 42:201–208, 2004. DOI: 10.1007/BF02344632 Cited on page(s) 39

[124] R. Zwiggelaar, S. M. Astley, C. R. M. Boggis, and C. J. Taylor. Linear structures in mammographic images: Detection and classification. *IEEE Transactions on Medical Imaging*, 23(9):1077–1086, September 2004. DOI: 10.1109/TMI.2004.828675 Cited on page(s) 41

[125] M. Samulski and N. Karssemeijer. Optimizing case-based detection performance in a multiview CAD system for mammography. *IEEE Transactions on Medical Imaging*, 30(4):1001–1009, April 2011. DOI: 10.1109/TMI.2011.2105886 Cited on page(s) 41

[126] G. S. Muralidhar, A. C. Bovik, J. D. Giese, M. P. Sampat, G. J. Whitman, T. M. Haygood, T. W. Stephens, and M. K. Markey. Snakules: a model-based active contour algorithm for the annotation of spicules on mammography. *IEEE Transactions of Medical Imaging*, 29(10):1768–1780, October 2010. DOI: 10.1109/TMI.2010.2052064 Cited on page(s) 41

[127] C. Evans, K. Yates, and M. Brady. Statistical characterization of normal curvilinear structures in mammograms. In H.-O. Peitgen, editor, *Proceedings of the 6th International Workshop on Digital Mammography (IWDM 2002)*, pages 285–291, Bremen, Germany, June 2002. Springer. Cited on page(s) 41

[128] L. C. C. Wai, M. Mellor, and M. Brady. A multi-resolution CLS detection algorithm for mammographic image analysis. In C. Barillot, D. R. Haynor, and P. Hellier, editors,

Lecture Notes in Computer Science, Proceedings of Medical Image Computing and Computer-Assisted Intervention (MICCAI 2004), pages 865–872, Berlin, Germany, 2004. Springer-Verlag. DOI: 10.1007/b100265 Cited on page(s) 41

[129] R. N. Dixon and C. J. Taylor. Automated asbestos fibre counting. *Institute of Physics Conference Series*, 44:178–185, 1979. Cited on page(s) 41

[130] R. Zwiggelaar, T. C. Parr, and C. J. Taylor. Finding orientated line patterns in digital mammographic images. In *Proceedings of the 7th British Machine Vision Conference*, pages 715–724, Edinburgh, UK, 1996. Cited on page(s) 41

[131] T. Lindeberg. Edge detection and ridge detection with automatic scale selection. *International Journal of Computer Vision*, 30(2):117–154, 1998. DOI: 10.1023/A:1008097225773 Cited on page(s) 41

[132] R. J. Ferrari, R. M. Rangayyan, R. A. Borges, and A. F. Frère. Segmentation of the fibroglandular disc in mammograms using Gaussian mixture modeling. *Medical and Biological Engineering and Computing*, 42:378–387, 2004. DOI: 10.1007/BF02344714 Cited on page(s) 41, 141

[133] M. Sonka, V. Hlavac, and R. Boyle. *Image Processing, Analysis and Machine Vision*. Chapman & Hall, London, UK, 1st edition, 1993. Cited on page(s) 42

[134] J. Canny. A computational approach to edge detection. *IEEE Transactions on Pattern Analysis and Machine Intelligence*, 8(6):679–698, 1986. DOI: 10.1109/TPAMI.1986.4767851 Cited on page(s) 42

[135] R. M. Rangayyan and F. J. Ayres. Detection of architectural distortion in mammograms using a shape-constrained phase portrait model. In H. U. Lemke, K. Inamura, K. Doi, M. W. Vannier, and A. G. Farman, editors, *Proceedings of the 20th International Congress and Exhibition on Computer Assisted Radiology and Surgery (CARS 2006)*, pages 334–336, Osaka, Japan, July 2006. Cited on page(s) 45

[136] K. M. Abadir and J. R. Magnus. *Matrix Algebra*. Cambridge University Press, New York, NY, 2005. Cited on page(s) 46

[137] H. Alto, R. M. Rangayyan, R. B. Paranjape, J. E. L. Desautels, and H. Bryant. An indexed atlas of digital mammograms for computer-aided diagnosis of breast cancer. *Annales des Télécommunications*, 58(5-6):820–835, 2003. DOI: 10.1007/BF03001532 Cited on page(s) 53

[138] Alberta Health Services, http://www.albertahealthservices.ca/services.asp?pid=service &rid=1002353. *Screen Test and the Alberta Breast Cancer Screening Program*, accessed November 2012. Cited on page(s) 53

[139] R. O. Duda, P. E. Hart, and D. G. Stork. *Pattern Classification*. Wiley-Interscience, New York, NY, 2nd edition, 2001. Cited on page(s) 63, 64, 74, 75, 76, 82

[140] L. Wang. Feature selection with kernel class separability. *IEEE Transactions on Pattern Analysis and Machine Intelligence*, 30(9):1534–1546, September 2008. DOI: 10.1109/TPAMI.2007.70799 Cited on page(s) 63, 68

[141] R. J. Nandi, A. K. Nandi, R. M. Rangayyan, and D. Scutt. Classification of breast masses in mammograms using genetic programming and feature selection. *Medical and Biological Engineering and Computing*, 44:683–694, 2006. DOI: 10.1007/s11517-006-0077-6 Cited on page(s) 63, 69

[142] B. Sahiner, H.-P. Chan, N. Petrick, R. F. Wagner, and L. Hadjiiski. Feature selection and classifier performance in computer-aided diagnosis: the effect of finite sample size. *Medical Physics*, 27(7):1509–1522, July 2000. DOI: 10.1118/1.599017 Cited on page(s) 63, 70, 84, 133

[143] J. H. Ware, F. Mosteller, F. Delgado, C. Donnelly, and J. A. Ingelfinger. P values. In J.C. Bailar III and F. Mosteller, editors, *Medical Uses of Statistics*, pages 181–200. NEJM Books, Boston, MA, second edition, 1992. Cited on page(s) 63, 64, 65, 66

[144] T. Mu, A. K. Nandi, and R. M. Rangayyan. Classification of breast masses using selected shape, edge-sharpness, and texture features with linear and kernel-based classifiers. *Journal of Digital Imaging*, 21(2):153–169, June 2008. DOI: 10.1007/s10278-007-9102-z Cited on page(s) 63

[145] W. H. Press, S. A. Teukolsky, W. T. Vetterling, and B. P. Flannery. *Numerical Recipes in C*. Cambridge University Press, New Delhi, India, second edition, 1988. Cited on page(s) 64, 65

[146] ROCKIT. Kurt Rossmann Laboratories for Radiologic Image Research. ROC Software. http://www-radiology.uchicago.edu/krl/roc_soft6.htm, accessed on March 15, 2012. Cited on page(s) 65, 66

[147] MATLAB, http://www.mathworks.com/products/matlab/. Cited on page(s) 66, 69

[148] T. Fawcett. An introduction to ROC analysis. *Pattern Recognition Letters*, 27:861–874, 2006. DOI: 10.1016/j.patrec.2005.10.010 Cited on page(s) 66

[149] H. L. Kundel, K. Berbaum, D. Dorfman, D. Gur, C. E. Metz, and R. G. Swensson, editors. *Journal of the ICRU. ICRU Report 79: Receiver Operating Characteristic Analysis in Medical Imaging*, volume 8(1), chapter 5. Extensions to Conventional ROC Methodology: LROC, FROC, and AFROC, pages 31–35. Oxford University Press, April 2008. Cited on page(s) 67

[150] D. P. Chakraborty. Validation and statistical power comparison of methods for analyzing free-response observer performance studies. *Academic Radiology*, 15(12):1554–1566, December 2008. DOI: 10.1016/j.acra.2008.07.018 Cited on page(s) 67, 125

[151] D. P. Chakraborty. Analysis of location specific observer performance data: validated extensions of the jackknife free-response (JAFROC) method. *Academic Radiology*, 13(10):1187–1193, October 2006. DOI: 10.1016/j.acra.2006.06.016 Cited on page(s) 67, 125

[152] K. Fukunaga. *Introduction to Statistical Pattern Recognition*. Academic Press Professional, Inc., San Diego, CA, 2 edition, 1990. Cited on page(s) 68, 131, 132

[153] F. L. Ramsey and D. W. Schafer. *The Statistical Sleuth: A Course in Methods of Data Analysis*. Duxbury Press, Belmont, CA, 1997. Cited on page(s) 69, 70

[154] P. Pudil, J. Novovičová, and J. Kittler. Floating search methods in feature selection. *Pattern Recognition Letters*, 15(11):1119—1125, 1994. DOI: 10.1016/0167-8655(94)90127-9 Cited on page(s) 69

[155] MATLAB Statistics Toolbox, Mathworks. Available online. http://www.mathworks.com/help/toolbox/stats/. Cited on page(s) 70, 71

[156] N. R. Draper and H. Smith. *Applied Regression Analysis*. Wiley-Interscience, Hoboken, NJ, 1998. Cited on page(s) 71

[157] R. Schalkoff. *Pattern Recognition: Statistical, Structural and Neural Approaches*. Wiley, New York, NY, 1992. Cited on page(s) 72, 73, 75, 78, 80

[158] S. Haykin. *Neural Networks: A Comprehensive Foundation*. Prentice Hall, Englewood Cliffs, NJ, second edition, 1999. Cited on page(s) 76, 77, 78, 79, 80, 81

[159] MATLAB Neural Network Toolbox, Mathworks. Available online. http://www.mathworks.com/products/neuralnet/. Cited on page(s) 79, 80, 81

[160] V. Vapnik. *Statistical Learning Theory*. Wiley, New York, NY, 1998. Cited on page(s) 81

[161] B. Schölkopf and A. J. Smola. *Learning with Kernels - Support Vector Machines, Regularization, Optimization, and Beyond*. MIT Press, Cambridge, MA, 2002. Cited on page(s)

[162] R. Fransens, J. D. Prins, and L. V. Gool. SVM-based nonparametric discriminant analysis, an application to face detection. In *Proceedings of the Ninth IEEE International Conference on Computer Vision (ICCV 2003)*, volume 2, pages 1289–1296. IEEE Computer Society, October 2003. DOI: 10.1109/ICCV.2003.1238639 Cited on page(s)

[163] C. J. C. Burges. A tutorial on support vector machines for pattern recognition. *Data Mining and Knowledge Discovery*, 2(2):121–167, 1998. DOI: 10.1023/A:1009715923555 Cited on page(s) 81

[164] T. Mu, A. K. Nandi, and R. M. Rangayyan. Classification of breast masses via nonlinear transformation of features based on a kernel matrix. *Medical and Biological Engineering and Computing*, 45(8):769–780, August 2007. DOI: 10.1007/s11517-007-0211-0 Cited on page(s) 82

[165] S. J. Raudys and A. K. Jain. Small sample size effects in statistical pattern recognition: Recommendations for practitioners. *IEEE Transactions on Pattern Analysis and Machine Intelligence*, 13(3):252–264, March 1991. DOI: 10.1109/34.75512 Cited on page(s) 83, 133

[166] G. C. Cawley and N. L. C. Talbot. Efficient leave-one-out cross-validation of kernel Fisher discriminant classifiers. *Pattern Recognition*, 36(11):2585–2592, November 2003. DOI: 10.1016/S0031-3203(03)00136-5 Cited on page(s) 83

[167] M. Yazdanpanah, L. Allard, L. G. Durand, and R. Guardo. Evaluation of Karhunen-Loève expansion for feature selection in computer-assisted classification of bioprosthetic heart-valve status. *Medical & Biological Engineering & Computing*, 37(4):504–10, July 1999. DOI: 10.1007/BF02513337 Cited on page(s) 83

[168] K. Fukunaga and R. R. Hayes. Effects of sample size in classifier design. *IEEE Transactions on Pattern Analysis and Machine Intelligence*, 11(8):873–885, August 1989. DOI: 10.1109/34.31448 Cited on page(s) 83, 133

[169] L. G. Durand, M. Blanchard, G. Cloutier, H. N. Sabbah, and P. D. Stein. Comparison of pattern recognition methods for computer-assisted classification of spectra of heart sounds in patients with a porcine bioprosthetic valve implanted in the mitral position. *IEEE Transactions on Biomedical Engineering*, 37(12):1121–1129, December 1990. DOI: 10.1109/10.64456 Cited on page(s) 83, 84

[170] D. H. Foley. Considerations of sample and feature size. *IEEE Transactions on Information Theory*, IT–18(5):618–626, September 1972. DOI: 10.1109/TIT.1972.1054863 Cited on page(s) 84

[171] T. W. Way, B. Sahiner, L. M. Hadjiiski, and H. P. Chan. Effect of finite sample size on feature selection and classification: a simulation study. *Medical Physics*, 37(2):907–920, February 2010. DOI: 10.1118/1.3284974 Cited on page(s) 84

[172] B. S. Sahiner, H. P. Chan, N. Petrick, M. A. Helvie, and M. M. Goodsitt. Computerized characterization of masses on mammograms: The rubber band straightening transform and texture analysis. *Medical Physics*, 25(4):516–526, 1998. DOI: 10.1118/1.598228 Cited on page(s) 89

[173] H. H. Barrett and K. J. Myers. *Foundations of Image Science*. Wiley, Hoboken, NJ, 2004. Cited on page(s) 91, 135

[174] P. Campisi and G. Scarano. A multiresolution approach for texture synthesis using the circular harmonic functions. *IEEE Transactions on Image Processing*, 11(1):37–51, January 2002. DOI: 10.1109/83.977881 Cited on page(s) 91, 135

[175] B. B. Mandelbrot. *The Fractal Geometry of Nature*. Freeman, San Francisco, CA, 1983. Cited on page(s) 92

[176] C. Fortin, R. Kumaresan, and W. Ohley. Fractal dimension in the analysis of medical images. *IEEE Engineering in Medicine and Biology Magazine*, 11:65–71, June, 1992. DOI: 10.1109/51.139039 Cited on page(s) 92

[177] R. M. Rangayyan and T. M. Nguyen. Fractal analysis of contours of breast masses in mammograms. *Journal of Digital Imaging*, 20(3):223–237, September 2007. DOI: 10.1007/s10278-006-0860-9 Cited on page(s) 92, 93

[178] T. M. Cabral and R. M. Rangayyan. *Fractal Analysis of Breast Masses in Mammograms*. Morgan & Claypool Publishers, 2012. DOI: 10.2200/S00453ED1V01Y201210BME046 Cited on page(s) 92, 93

[179] Q. Huang, J. R. Lorch, and R. C. Dubes. Can the fractal dimension of images be measured? *Pattern Recognition*, 27:1569–1579, 1994. DOI: 10.1016/0031-3203(94)90112-0 Cited on page(s) 92

[180] H. E. Schepers, J. H. G. M. van Beek, and J. B. Bassingthwaighte. Four methods to estimate the fractal dimension from self-affine signals. *IEEE Engineering in Medicine and Biology Magazine*, 11:57–64, June 1992. DOI: 10.1109/51.139038 Cited on page(s) 92

[181] P. Bak, C. Tang, and K. Wiesenfeld. Self-organized criticality: An explanation of 1/f noise. *The American Physical Society*, 59:381–384, 1987. DOI: 10.1103/PhysRevLett.59.381 Cited on page(s) 92, 93

[182] S. B. Lowen and M. C. Teich. Fractal renewal processes generate 1/f noise. *The American Physical Society*, 47:992–1001, 1993. DOI: 10.1103/PhysRevE.47.992 Cited on page(s) 92

[183] P. De Los Rios and Y.-C. Zhang. Universal / noise from dissipative self-organized criticality models. *The American Physical Society*, 82:472–475, 1999. DOI: 10.1103/PhysRevLett.82.472 Cited on page(s) 92

[184] V. A. Billock, G. C. De Guzman, and J. A. S. Kelso. Fractal time and 1/f spectra in dynamic images and human vision. *Physica D: Nonlinear Phenomena*, 148:136–146, 2001. DOI: 10.1016/S0167-2789(00)00174-3 Cited on page(s) 93

[185] T. Stŏsić and B. D. Stŏsić. Multifractal analysis of human retinal vessels. *IEEE Transactions on Medical Imaging*, 25:1101–1107, 2006. DOI: 10.1109/TMI.2006.879316 Cited on page(s) 93

[186] H.-O. Peitgen, H. Jürgens, and D. Saupe. *Chaos and Fractals: New Frontiers of Science*. Springer, New York, NY, second edition, 2004. Cited on page(s) 93, 141

[187] S. Banik, R. M. Rangayyan, and J. E. L. Desautels. Detection of architectural distortion in prior mammograms of interval cancer using measures of angular spread and Tsallis entropy. In *Proceedings of the 25th International Congress and Exhibition: Computer Assisted Radiology and Surgery*, volume 6, pages S188—S189, Berlin, Germany, June 2011. DOI: 10.1117/12.877587 Cited on page(s) 108, 121, 135

[188] R. M. Rangayyan, S. Banik, and J. E. L. Desautels. Detection of architectural distortion in prior mammograms using measures of angular distribution. In R. M. Summers and B. van Ginneken, editors, *Proceedings of SPIE Medical Imaging 2011: Computer Aided Diagnosis*, volume 7963, pages 796308: 1–9, Orlando, FL, February 2011. Cited on page(s) 108, 121, 135

[189] N. R. Mudigonda, R. M. Rangayyan, and J. E. L. Desautels. Detection of breast masses in mammograms by density slicing and texture flow-field analysis. *IEEE Transactions on Medical Imaging*, 20(12):1215–1227, 2001. DOI: 10.1109/42.974917 Cited on page(s) 110, 111

[190] R. M. Rangayyan, S. Banik, and J. E. L. Desautels. Detection of architectural distortion in prior mammograms using measures of angular dispersion. In *Proceedings of the 2012 IEEE International Symposium on Medical Measurements and Applications (MeMeA)*, pages 87–90, Budapest, Hungary, May 2012. DOI: 10.1109/MeMeA.2012.6226626 Cited on page(s) 113, 135

[191] C. E. Shannon. A mathematical theory of communication. *Bell System Technical Journal*, 27:379–423, 623–656, 1948. DOI: 10.1145/584091.584093 Cited on page(s) 113

[192] A. Rényi. On measures of entropy and information. In *Proceedings of the 4th Berkeley Symposium on Mathematics, Statistics and Probability*, volume 1, pages 547–561, Berkeley, CA, 1961. University of California Press. Cited on page(s) 113, 114, 115

[193] M. Masi. A step beyond Tsallis and Rényi entropies. *Physics Letters A*, 338(3–5):217–224, May 2005. DOI: 10.1016/j.physleta.2005.01.094 Cited on page(s) 114, 115

[194] C. Tsallis. Possible generalization of Boltzmann-Gibbs statistics. *Journal of Statistical Physics*, 52(1):479–487, July 1988. DOI: 10.1007/BF01016429 Cited on page(s) 114

[195] R. S. Johal and U. Tirnakli. Tsallis versus Renyi entropic form for systems with q-exponential behaviour: the case of dissipative maps. *Physica A*, 331(3–4):487—496, January 2004. DOI: 10.1016/j.physa.2003.09.064 Cited on page(s) 114

[196] D. Zhang, X. Jia, H. Ding, D. Ye, and N. V. Thakor. Application of Tsallis entropy to EEG: Quantifying the presence of burst suppression after asphyxial cardiac arrest in rats. *IEEE Transactions of Biomedical Engineering*, 57(4):867–874, April 2010. DOI: 10.1109/TBME.2009.2029082 Cited on page(s) 114, 115

[197] P. S. Rodrigues, G. A. Giraldi, R.-F. Chang, and J. S. Suri. Non-extensive entropy for CAD systems of breast cancer images. In *The 19th Brazilian Symposium on Computer Graphics and Image Processing*, pages 121–128, 2006. DOI: 10.1109/SIBGRAPI.2006.31 Cited on page(s)

[198] P. S. Rodrigues and G. A. Giraldi. Computing the q-index for Tsallis nonextensive image segmentation. In *Brazilian Symposium on Computer Graphics and Image Processing*, pages 232–237, 2009. DOI: 10.1109/SIBGRAPI.2009.23 Cited on page(s) 115

[199] W. Kinsner. A unified approach to fractal dimensions. In *Proceedings of the Fourth IEEE International Conference on Cognitive Informatics (ICCI)*, pages 58–72, Irvine, CA, August 2005. IEEE Computer Society. DOI: 10.1109/COGINF.2005.1532616 Cited on page(s) 115, 141

[200] S. Gabarda and G. Cristóbal. Discrimination of isotrigon textures using the Rényi entropy of Allan variances. *Journal of the Optical Society of America A*, 25(9):2309–2319, September 2008. DOI: 10.1364/JOSAA.25.002309 Cited on page(s) 115

[201] Y. Li, X. Fan, and G. Li. Image segmentation based on Tsallis-entropy and Renyi-entropy and their comparison. In *IEEE International Conference on Industrial Informatics*, pages 943–948, Singapore, August 2006. DOI: 10.1109/INDIN.2006.275704 Cited on page(s) 115

[202] P. Sahoo, C. Wilkins, and J. Yeager. Threshold selection using Renyi's entropy. *Pattern Recognition*, 30(1):71–84, January 1997. DOI: 10.1016/S0031-3203(96)00065-9 Cited on page(s) 115

[203] S. Banik, R. M. Rangayyan, and J. E. L. Desautels. Rényi entropy of angular spread for detection of architectural distortion in prior mammograms. In *Proceedings of the 2011 IEEE International Symposium on Medical Measurements and Applications (MeMeA)*, pages 609—612, Bari, Italy, May 2011. DOI: 10.1109/MeMeA.2011.5966645 Cited on page(s) 121, 135

[204] H.-P. Chan, B. Sahiner, R. F. Wagner, and N. Petrick. Classifier design for computer-aided diagnosis: Effects of finite sample size on the mean performance of classical and neural network classifiers. *Medical Physics*, 26(12):2654–2668, December 1999. DOI: 10.1118/1.598805 Cited on page(s) 131, 132

[205] N. A. Obuchowski. Nonparametric analysis of clustered ROC curve data. *Biometrics*, 53(2):567–578, June 1997. DOI: 10.2307/2533958 Cited on page(s) 136

[206] F. W. Samuelson, N. Petrick, and S. Paquerault. Advantages and examples of re-sampling for CAD evaluation. In *4th IEEE International Symposium on Biomedical Imaging: From Nano to Macro (ISBI 2007)*, pages 492–495, Arlington, VA, April 2007. DOI: 10.1109/ISBI.2007.356896 Cited on page(s) 137

[207] K. Y. Liu, M. R. Smith, E. Fear, and R. M. Rangayyan. Evaluation and amelioration of computer-aided diagnosis with artificial neural networks utilizing small-sized sample sets.

Biomedical Signal Processing and Control, 2012. In press. DOI: 10.1016/j.bspc.2012.11.001
Cited on page(s)

[208] K. Y. Liu, M. R. Smith, and R. M. Rangayyan. The application of Efron's bootstrap methods in validating feature classification using artificial neural networks for the analysis of mammographic masses. In *Proceedings of the 26th Annual International Conference of the IEEE Engineering in Medicine and Biology Society*, pages 1553–1556, San Francisco, CA, September 2004. DOI: 10.1109/IEMBS.2004.1403474 Cited on page(s) 137

[209] J. Li, W.-Y. Yau, and H. Wang. Constrained nonlinear models of fingerprint orientations with prediction. *Pattern Recognition*, 39(1):102–114, January 2006.
DOI: 10.1016/j.patcog.2005.08.010 Cited on page(s) 141

[210] R. M. Ford and R. N. Strickland. Nonlinear phase portrait models for oriented textures. In *Proceedings of Computer Vision and Pattern Recognition (CVPR 1993)*, pages 644–645, New York, NY, June 1993. IEEE Computer Society. DOI: 10.1109/CVPR.1993.341047 Cited on page(s) 141

[211] R. J. Ferrari, R. M. Rangayyan, J. E. L. Desautels, R. A. Borges, and A. F. Frère. Automatic identification of the pectoral muscle in mammograms. *IEEE Transactions on Medical Imaging*, 23:232–245, 2004. DOI: 10.1109/TMI.2003.823062 Cited on page(s) 141

[212] M. Sandri. Numerical calculation of Lyapunov exponents. *The Mathematica Journal*, 6(3):78–84, 1996. Cited on page(s) 141

[213] H. Kantz, J. Kurths, and G. Mayer-Kress, editors. *Nonlinear Analysis of Physiological Data*. Springer, Berlin, Germany, April 1998. DOI: 10.1007/978-3-642-71949-3 Cited on page(s) 141

[214] W. Klonowski. Signal and image analysis using chaos theory and fractal geometry. *Machine Graphics & Vision*, 9:403–431, May 2000. Cited on page(s) 141

AUTHORS' BIOGRAPHIES

Shantanu Banik received his Ph.D. in 2011 and M.Sc. in 2008 from the Department of Electrical and Computer Engineering, University of Calgary, Calgary, Alberta, Canada, and his B.Sc. in 2005 in Electrical and Electronic Engineering from the Bangladesh University of Engineering and Technology (BUET), Dhaka, Bangladesh. His Ph.D. thesis was on the problem of detection of architectural distortion in prior mammograms to aid the process of early detection of breast cancer. His research interests include medical signal and image processing and analysis, development of computer-aided diagnosis (CAD) techniques for the detection of cancer, landmarking and segmentation of medical images, pattern recognition and classification, medical imaging, and automatic segmentation and analysis of tumors. He has coauthored several journal papers, a number of conference papers, three book chapters, and a book titled *Landmarking and Segmentation of 3D CT Images* (Morgan & Claypool, 2009). He is currently writing two more books on image processing and biomedical applications. He received many awards and scholarships as a graduate student at the University of Calgary, including the Institute of Cancer Research (ICR), Canada Publication Prize for significant contribution on cancer research; Natural Sciences and Engineering Research Council (NSERC) of Canada and Collaborative Research and Training Experience (CREATE) postdoctoral fellowship; J. B. Hyne Research Innovation Award for outstanding research activity at the University of Calgary; Robert B. Paugh Memorial Award; Graduate Student Productivity Award; the Queen Elizabeth II Graduate (Doctoral) Scholarship; the Graduate Faculty Council Scholarship (Doctoral); the University Technologies International Inc. (UTI) Fellowship; the University of Calgary Alumni Association Graduate Scholarship; and the Schulich School of Engineering Teaching Assistant Excellence Award. He is currently working as a Research and Developement Engineer at the Circle Cardiovascular Imaging, Calgary, Alberta, Canada.

Rangaraj Mandayam Rangayyan is a Professor with the Department of Electrical and Computer Engineering, and an Adjunct Professor of Surgery and Radiology, at the University of Calgary, Calgary, Alberta, Canada. He received a Bachelor of Engineering degree in Electronics and Communication in 1976 from the University of Mysore at the People's Education Society College of Engineering, Mandya, Karnataka, India, and a Ph.D. in Electrical Engineering from the Indian Institute of Science, Bangalore, Karnataka, India, in 1980. His research interests are in the areas of digital signal and image processing, biomedical signal analysis, biomedical image analysis, and computer-aided diagnosis. He has published more than 150 papers in journals and 250 papers in proceedings of conferences. His research productivity was recognized with the 1997 and 2001 Research Excellence Awards of the Department of Electrical and Computer Engineering, the 1997 Research Award of the Faculty of Engineering, and by appointment as a "University Professor" in 2003, at the University of Calgary. He is the author of two textbooks: *Biomedical Signal Analysis* (IEEE/ Wiley, 2002) and *Biomedical Image Analysis* (CRC, 2005). He has coauthored and coedited several other books, including *Color Image Processing with Biomedical Applications* (SPIE, 2011). He was recognized by the IEEE with the award of the Third Millennium Medal in 2000, and was elected as a Fellow of the IEEE in 2001, Fellow of the Engineering Institute of Canada in 2002, Fellow of the American Institute for Medical and Biological Engineering in 2003, Fellow of SPIE: the International Society for Optical Engineering in 2003, Fellow of the Society for Imaging Informatics in Medicine in 2007, Fellow of the Canadian Medical and Biological Engineering Society in 2007, and Fellow of the Canadian Academy of Engineering in 2009. He has been awarded the Killam Resident Fellowship thrice (1998, 2002, and 2007) in support of his book-writing projects.

Dr. J. E. Leo Desautels obtained his M.D. from the University of Ottawa in 1955, and completed post-graduate training in radiology at the Henry Ford Hospital, Detroit, MI. He was a Staff Radiologist at the Foothills Hospital and a Clinical Professor with the Faculty of Medicine, the University of Calgary, Calgary, AB, Canada, from 1970 to 1994. He served as a Reference Radiologist to the Alberta Program for the Early Detection of Breast Cancer until 2007. He is an Adjunct Professor of Electrical and Computer Engineering at the University of Calgary. He is interested in computer applications in mammography.

Printed in the United States
by Baker & Taylor Publisher Services